Physical Geography

Process and System

Ken Briggs

HODDER AND STOUGHTON
LONDON SYDNEY AUCKLAND TORONTO

British Library Cataloguing in Publication Data

Briggs, K.
Physical geography: process and system.
1. Physical geography—Text-books—1945–
I. Title
910°.02 GB55

ISBN 0 340 35951 X

First published 1985
Fifth Impression 1989

Printed in Great Britain for the educational publishing division of Hodder and Stoughton Ltd, Mill Road, Dunton Green, Sevenoaks, Kent by Thomson Litho Ltd, East Kilbride

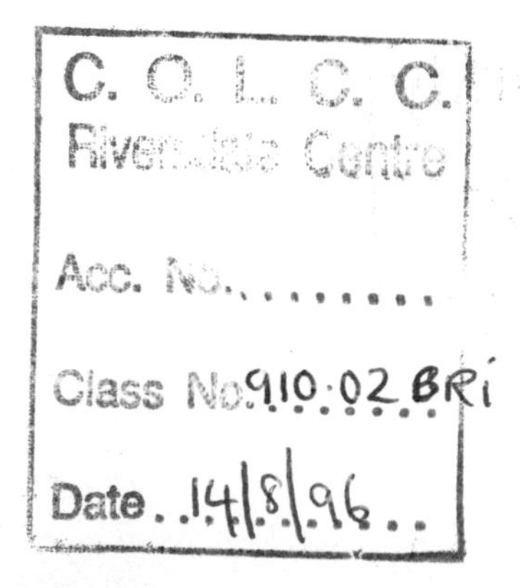

Contents

Preface

Although this book is divided into four separate sections, in the traditional manner, emphasis is placed throughout on the interrelationships between the various elements of the physical environment. Each successive section tends to be dependent upon earlier sections. Climate is shown to be influenced by geomorphology, vegetation by climate, and soils by all of these. But emphasis is also placed on the reverse linkages. Geomorphology is influenced by climate, climate by vegetation, and vegetation by soils.

Explanations are given of the operation of various physical processes and the ways in which such processes are related to other processes and integrated into physical systems at various scales. The underlying theme is that the physical environment can be viewed as being composed of hierarchies of interlocking and interacting systems. Despite this emphasis on process and system, descriptive aspects of physical geography are not neglected. Where appropriate, the relationships between physical systems and man are explained.

The exercises at the end of each major section are designed to consolidate learning and also to provide practice in the writing of essays and the answering of the data-response questions which some examining boards have adopted.

K. Briggs

Acknowledgements

The author and publishers are grateful to the following, who supplied photographs for use in this book: Patrick Bailey (p. 115); Eric Kaye (p. 15 bottom and 47 top); the South African Tourist Board, SATOUR, (p. 148); the Soil Survey of England and Wales (p. 162); Topham p. 145 bottom and cover. The remaining photographs were taken by the author.

Figure 2.27 is adapted from 'Winter Temperatures of a Mid-Latitude Desert Mountain Range', R. F. Logan, *Geographical Review* April 1961.
Data used in Figure 2.29 are derived from 'Climatic Types and the Variation of Moisture Regions in Turkey', Erinç, Sirri, *Geographical Review* April 1950.

1 Geomorphology

1.1 Major relief features of the earth

Relief features are irregularities in the solid surface of the earth. They include the mountains, hills, valleys and plains that we are able to see every day. But only 29 % of the earth's solid surface is above sea level (Fig. 1.1) and therefore readily visible. By far the greater part lies below sea level and is therefore hidden from view. Figure 1.1 shows how the earth's solid surface is concentrated at particular heights and depths. Over half of it lies between 3 km and 6 km below sea level. Only slightly more than 8 % is situated more than 1 km (1000 m) above sea level.

MAJOR RELIEF FEATURES LOCATED ABOVE SEA LEVEL

The surfaces of the earth's land areas have remarkably little relief. By far the greatest amount of land lies below an altitude of 1000 m (Fig. 1.1), and it exists in the form of coastal plains and low hills and comparatively shallow valleys. Hardly any part of the land surface rises to an altitude of 4000 m although Mt Everest reaches nearly 9000 m above sea level. Figure 1.2 shows the distribution of ancient shields and ranges of fold mountains of different ages.

Ancient shields

Shields usually have very little relief. They usually form lowlands or low plateaux. The Canadian Shield and western Australia are mostly below 500 m and the Baltic Shield rarely rises above 200 m. Africa, on the other hand, has large areas in the east and south that reach a height of over 1000 m and, unusually, has a very high mountain, Kilimanjaro, which soars to 5894 m above sea level. Near this mountain, in East Africa, is another of the earth's major relief features, the Great East African Rift Valley, a deep, faulted valley some 3000 km long. The shield areas of Brazil and Venezuela in South America rise to

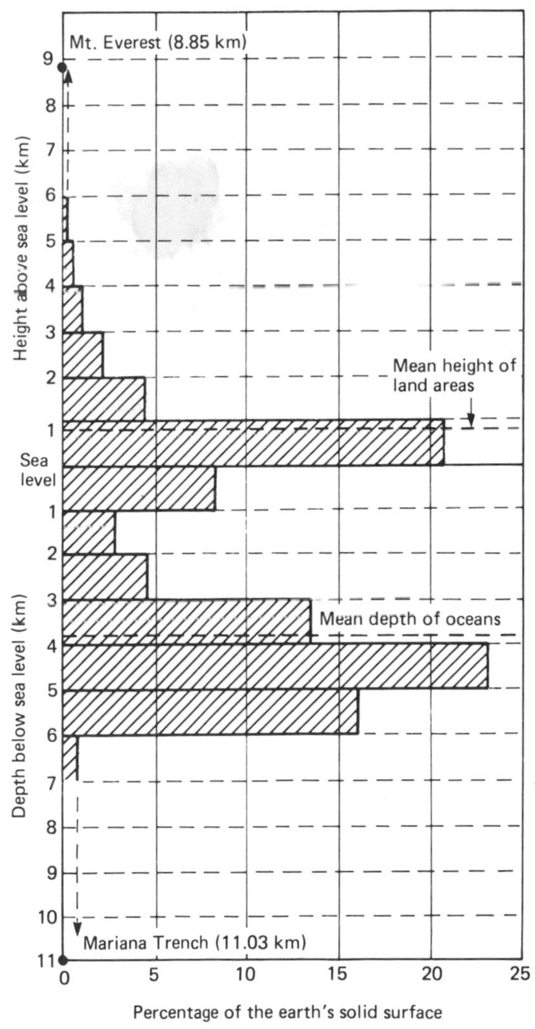

Fig. 1.1 Altitude distribution of the earth's solid surface

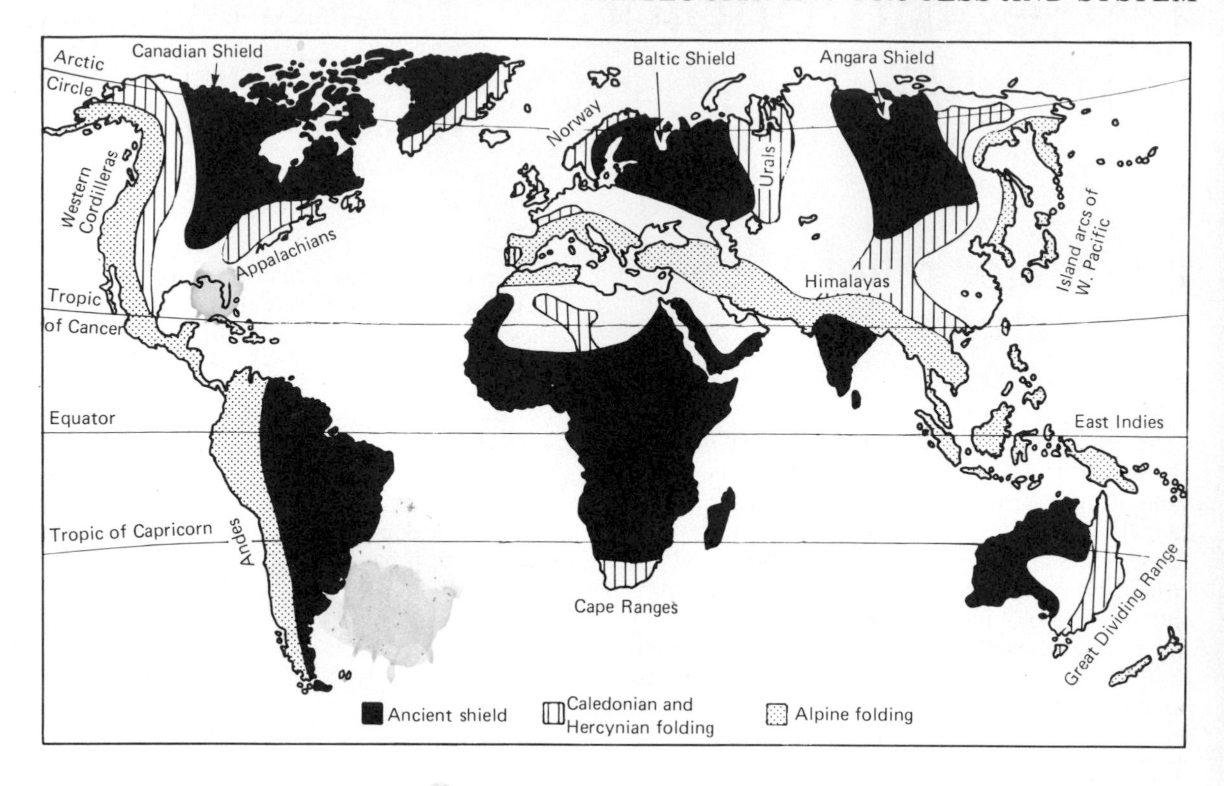

Fig. 1.2 The distribution of ancient shields and ranges of fold mountains of different ages

over 2000 m and have considerable variations in relief. Most of the continent of Antarctica is also a shield. A number of islands such as Greenland, Madagascar and Sri Lanka, are detached fragments of the major shield areas.

These shields are composed of extremely ancient rocks (Pre-Cambrian) older than 600 million years. During Pre-Cambrian times they suffered folding and igneous activity that created huge masses of igneous rock (batholiths). The ancient fold mountains have now been worn down by long ages of erosion, and the shields are partly covered by newer sedimentary rocks. The shields are regarded as comparatively stable parts of the earth's surface. They have very few earthquakes and little volcanic activity except in East Africa (Fig. 1.3).

Fold mountains

In contrast to the shields, fold mountains tend to form long, narrow belts stretching for considerable distances. Alpine fold mountains extend, for example, from Spain to Malaysia and from Alaska to southern Chile. The Andes range alone is over 7000 km in length, approximately the distance from London to Bombay (Fig. 1.2). In these ranges the rocks have been compressed into folds and uplifted. In some of the areas of Alpine folding (Fig. 1.2) frequent earthquakes (Fig. 1.3) show that considerable earth movements are still in progress. There is also considerable volcanic activity in most of the Alpine areas, especially in southern Italy, central America, the Andes and the Cascades (Mt St Helens), but not in the Himalayas. Figures 1.3 and 1.4 show that the ranges of Alpine fold mountains pass laterally into the 'island arcs' of the Pacific Ocean. These islands, which also experience frequent earthquake and volcanic activity, link the south-eastern end of the Alpine-Himalayan fold mountain belt with New Zealand to the south-east and with Alaska to the north-east, via Japan. This forms what has been called the Pacific 'ring of fire'.

The greatest of the Alpine ranges is the Himalayas-Karakoram range, north of India, which possesses the 52 highest mountain peaks in

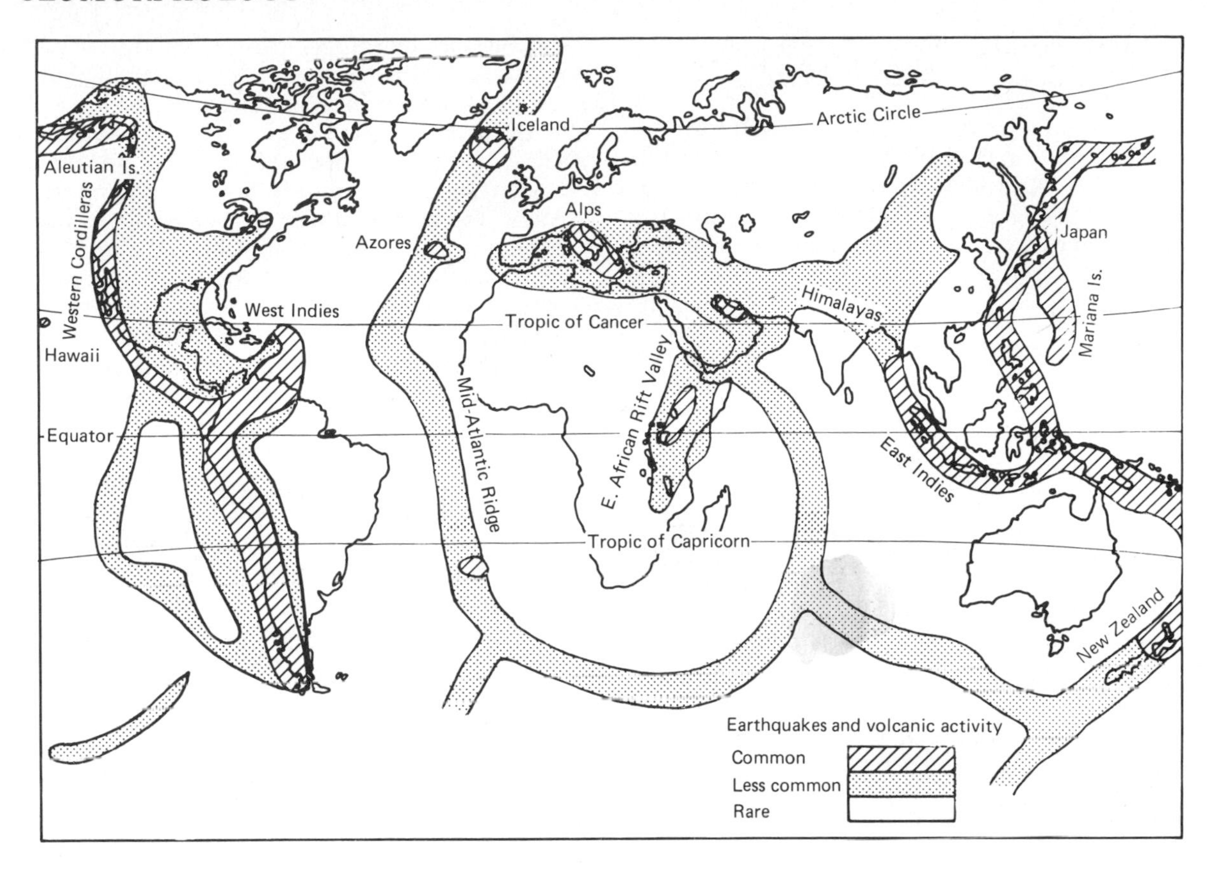

Fig. 1.3 Areas of volcanic activity and earthquakes

the world, including Everest in the Himalayas (8848 m) and K2 in the Karakoram (8610 m). Outside this area the highest mountain is Aconcagua (6960 m) in the Andes of Argentina. Mt McKinley (6194 m), in the Alaska Range, is the highest mountain in North America. By comparison the Alps themselves are fairly small, Mont Blanc reaching only 4807 m. In fact, the much less well known Caucasus range between the Black Sea and the Caspian Sea has 18 peaks all of which are higher than Mont Blanc, and Elbrus reaches 5633 m. One of the most striking examples of Alpine relief exists near Antofagasta in Chile where an Andean summit of 6723 m exists side by side with an ocean depth of 7973 m, giving a total relief variation of 14 696 m.

The Alpine fold mountains have been created in the last 40 million years and in fact mountain building is still in progress, as the frequency of earthquakes indicates. Older fold mountain ranges (Caledonian and Hercynian) often lie between the Alpine ranges and the ancient shields (Fig. 1.2). Folding took place between 500 and 250 million years ago, so their relief is generally more subdued. These uplands are frequently plateaux from which rise hills or low mountains. Such areas are central Asia north of the Alpine ranges, with summits rising to 3000–4000 m, and central Spain (2000–3000 m). The highest summit in the Australian Great Dividing Range is Kosciusko (2228 m). Ben Nevis in Scotland reaches 1344 m. A few areas have rather more vigorous relief features. These include the Cape Ranges of South Africa, the Urals, and Norway. In Norway summits only rise to about 2500 m but the landscape is strikingly dissected by glaciated valleys. Earthquakes and volcanoes are rare in these older fold mountain ranges.

MAJOR RELIEF FEATURES LOCATED BELOW SEA LEVEL

The most striking fact is that almost 60% of the area of the ocean floor lies between 3 km and 6 km deep. Also a considerable proportion is less than 1 km deep. The depth of the deepest part of the oceans is considerably greater than the height of Mt Everest (Fig. 1.1).

The continental shelf

The continental shelf consists of large areas of shallow sea floor (less than 180 m) that fringe some of the continents. Considerable areas of continental shelf exist off north-west Europe, eastern North America, eastern and south-east Asia, and to the east of Argentina. Few areas of continental shelf exist off the coasts of Africa or the west coasts of North and South America. On the continental shelf of north-west Europe the average depth of the North Sea is 90 m, of the Irish Sea is 60 m, and of the English Channel is only 54 m. At the outer edge of the continental shelf there is a steep descent to the ocean depths (the continental slope). Westwards from Land's End, for a distance of almost 400 km, the sea is shallower than 200 m; then in the next 30 km the sea bed descends to a depth of 1000 m. Many islands are situated on continental shelves. Examples include the British Isles, Tasmania, Borneo, and the Falkland Islands. A very small change of sea level would join these islands to their neighbouring mainland. Some seas which one might expect to be underlain by a continental shelf are quite deep. The Gulf of Mexico has an average depth of 1500 m which is similar to that of the Mediterranean. The outer edge of the continental shelf gives a more accurate indication of the shape of the continental masses than does the actual coastline.

The abyssal plains

Enormous areas of the world's oceans are underlain by vast, relatively level areas at depths of approximately 5 km. The North Atlantic, for example, has a relatively smooth floor at a depth of about 5500 m. Occasionally seamounts rise from these abyssal plains. The Hawaiian islands, for example, a number of active and extinct volcanoes, rise from such depths to reach a height of about 4000 m above sea level, giving a total vertical extent similar to the height of Mt Everest.

The mid-oceanic ridges

In places the abyssal plains are interrupted by long ranges of submarine mountains of a similar height to the Alps or the Rockies of North America. These mountains are not intensely folded or eroded, so they often form broad, smooth ridges. Along the summit there is usually a rift valley which may be about 30 km wide and about 2 km deep. Such a mid-oceanic ridge exists along the centre of the Atlantic and Indian Oceans and approximately midway between Antarctica and the continents of Africa and Australia (Fig. 1.4). Volcanic and earthquake activity occurs at intervals along it (Fig. 1.3). Volcanic islands include St. Helena, the Azores and Iceland which is the largest island in the world that has been built up entirely by volcanic activity. In 1963 a completely new island, Surtsey, appeared near Iceland. In the Pacific Ocean the 'mid-oceanic ridge' is not in the centre of the ocean but is much nearer to the Americas than to Asia and Australia. In fact it reaches the west coast of North America (Fig. 1.4).

The deep-sea trenches

The deepest parts of the world's oceans occur in the western Pacific in association with the island arcs that exist there (Fig. 1.4). Here, long, narrow depressions, well over 10 km deep, are found in the ocean floor, generally on the outer (convex) sides of the island arcs. The Tonga-Kermadec Trench, north of New Zealand, and the Mariana Trench, south of Japan, are particularly deep. These trenches continue northwards, past the Kuril Islands to the Aleutians, where the trench is not so deep (about 6.7 km). Deep-sea trenches are also found to the south of Java, along the west coast of Central America and South America, and to the north of the island of Puerto Rico in the West Indies (Fig. 1.4). The deep-sea trenches are not only very deep, but they are also very long and narrow. Both the Peru-Chile Trench and the Aleutian Trench are over 3000 km long. Figure 1.3 shows that the deep-sea trenches are closely associated with earthquakes and volcanic activity.

PLATE TECTONICS

In the early twentieth century Alfred Wegener attempted to explain some of the earth's major relief features by suggesting that the continents had been able to move about over the earth's surface. This idea was thought to be supported by,

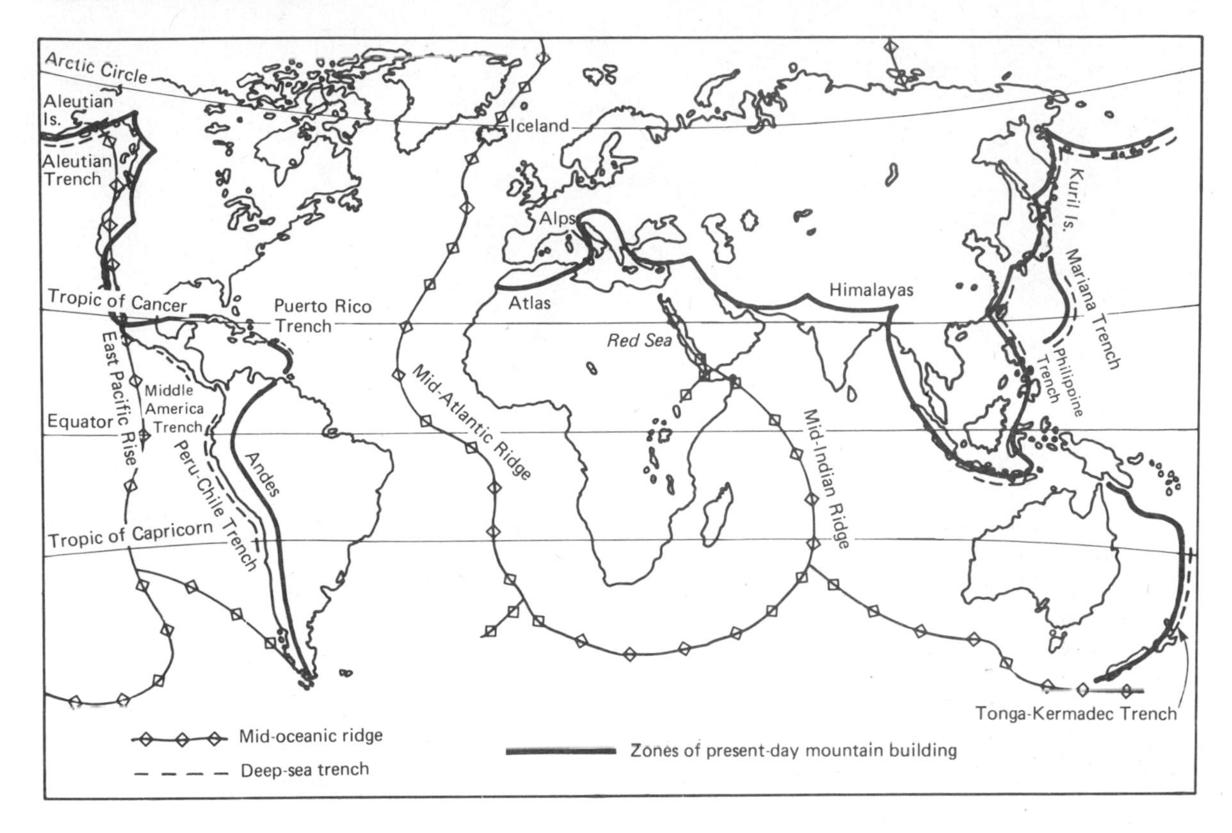

Fig. 1.4 Ridges, trenches, and zones of present-day mountain building

amongst other things, the correspondence in shape of the coastlines to the west and east of the Atlantic Ocean. If North and South America were moved to the east they would fit quite snugly against the west coasts of Europe and Africa. This idea was treated with scepticism mainly because it was not possible to envisage a force that was capable of moving whole continents as if they were huge rafts. Now Wegener's theory of 'continental drift' has been replaced by the theory of 'plate tectonics'. The main difference is that, instead of regarding the earth as a number of continents separated by oceans, plate tectonics envisages a number of crustal 'plates' (Fig. 1.5) which include both continental land areas and ocean floors.

Characteristics of crustal plates

The earth's crust varies in thickness from about 5 km under the oceans to about 40 km under the continents. The thin crust of the ocean floors is composed of basalt. This basaltic crust also extends beneath the continents but, here, it is overlain by a much thicker mass of granitic rock of lower density than the basalt (Fig. 1.6). Thus, the real edge of the continent is the 'continental slope' that descends steeply from the continental shelf to the ocean depths. Figure 1.6 makes it clear that a single crustal plate may include both a continent and an area of ocean floor. Most of the plates shown in Figure 1.5 are of this type, but the Pacific plate is entirely oceanic.

The movement of crustal plates is thought to be the result of convection currents in the upper part of the mantle, probably in a weak part of the mantle known as the asthenosphere. Plate movement is always away from the mid-oceanic ridges, as can be seen in Figure 1.5. New oceanic crust is created at the mid-oceanic ridges by the upwelling of basalt to replace oceanic crust that has 'spread' laterally (Fig. 1.7). The mid-oceanic ridges are therefore referred to as 'constructive plate margins'. If new crust is being created at the mid-

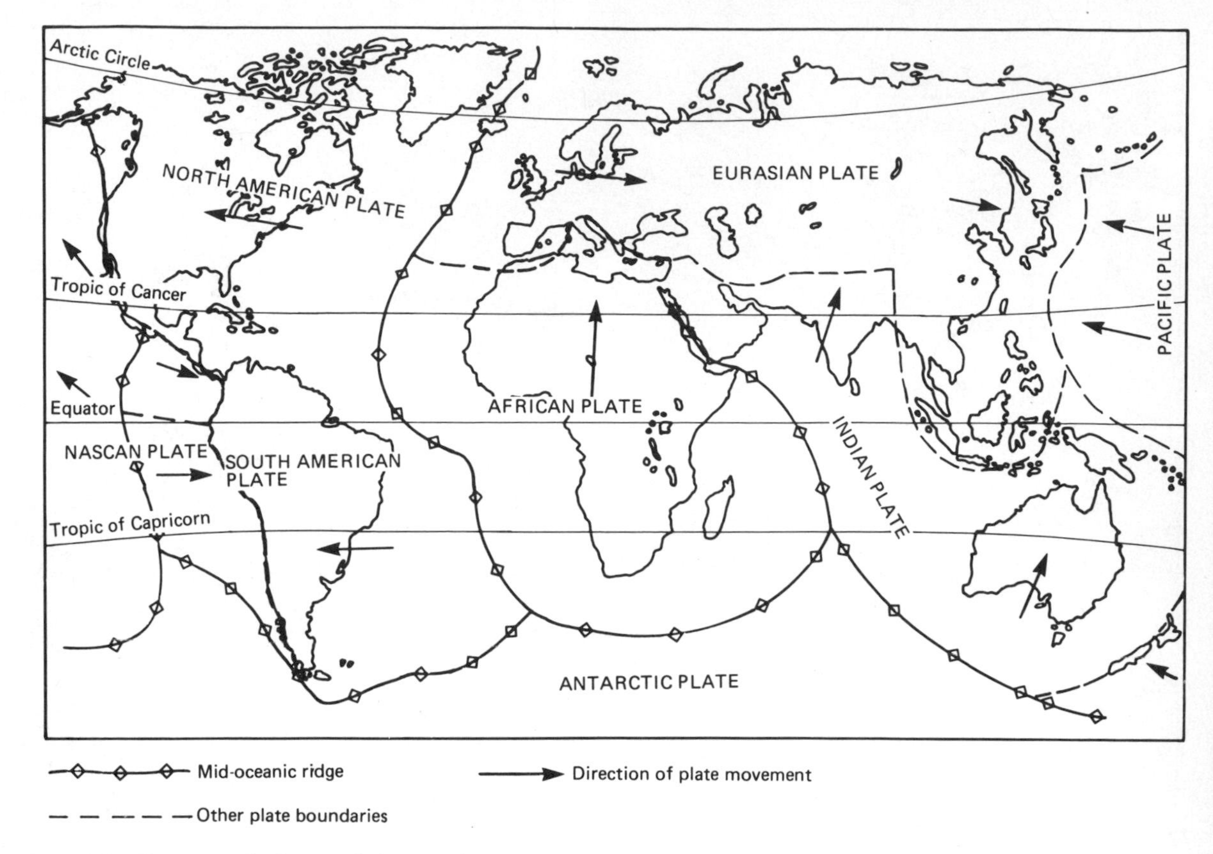

Fig. 1.5 The crustal plates of the earth

oceanic ridges then it is inevitable that there must be locations where oceanic crust is being destroyed. Such locations are called 'destructive plate margins'. They probably occur where there are downward convection currents in the mantle. At such a plate margin, the denser oceanic crust (basaltic) may sink under lighter (granitic) continental crust (Fig. 1.7). This results in the formation of island arcs and deep-sea trenches. In eastern Asia the westward moving Pacific plate meets the eastward moving Eurasian plate and is subducted beneath it, producing island arcs, the deepest parts of the world's oceans, and a great concentration of volcanoes and deep-focus earthquakes. Rather similar situations are found in the East Indies where the Indian plate meets the Eurasian plate and on the west coast of South America where the Nascan plate meets the South American plate (Fig. 1.5). In this latter example there is a prominent deep-sea trench, but there are no island arcs (Fig. 1.4).

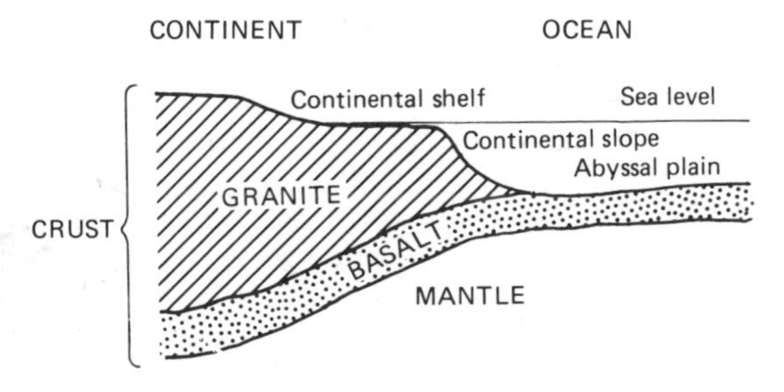

Fig. 1.6 Continental and oceanic crust

The creation of fold mountain systems

Various ranges of fold mountains have been created in the past, usually by collisions between two or more continental plates which obliterated a pre-existing sea so that its sediments were compressed to form fold mountains. In the fairly recent geological past a sea (Tethys) separated Laurasia (the northern continents of North

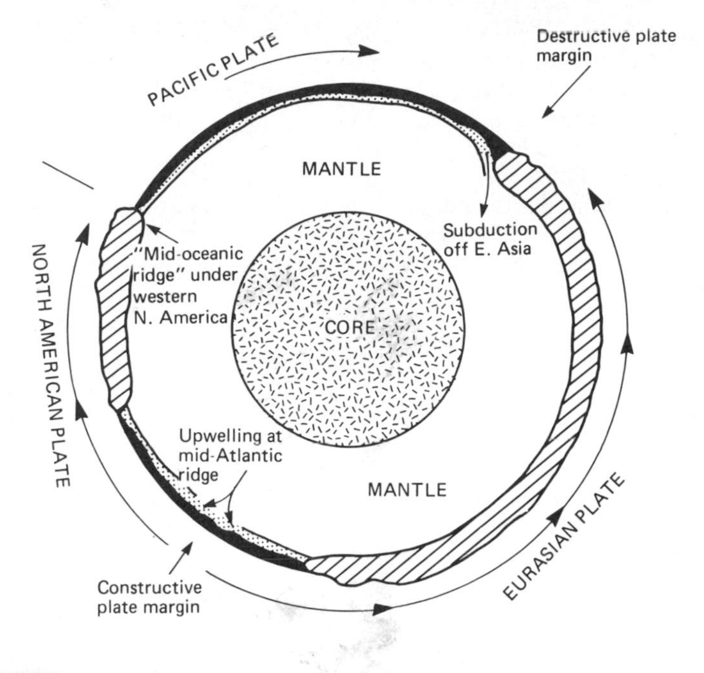

Fig. 1.7 Generalized section of the earth at about latitude 40° N showing plate tectonic processes (not to scale)

America, Europe, and Asia) from Gondwana (the southern continents of South America, Africa, Australia, Antarctica, and India). Africa and India then moved north and folded the Tethys sediments against Europe and Asia (Fig. 1.5). The Alpine fold mountains of the west of North and South America, however, were created by these continents colliding with the Pacific plate to the west.

1.2 Weathering, mass wasting and slope processes

WEATHERING

Weathering includes all the processes, both physical and chemical, that result in the breakdown and decomposition of rock at or near the earth's surface. By weathering, massive coherent rock is changed into an unconsolidated mantle of rock material which overlies so far unweathered bedrock, and is referred to as the waste mantle or the *regolith*. This may eventually develop into soil. By weathering, many types of rocks originally created under conditions of high temperature and high pressure are transformed into materials which are more in equilibrium with conditions at or near the earth's surface.

Physical weathering

Physical weathering is the breakdown of rock by mechanical means so that it disintegrates into blocks, fragments and grains. It is frequently the result of climatic influences. Three main types are usually recognized: granular disintegration, exfoliation and block separation. *Granular disintegration* occurs when rock divides into tiny grains to form sand or very fine gravel. *Exfoliation* occurs when thin, curved sheets fall away from the rock surface. The break-up of rock into large fragments related to bedding planes or joints is known as *block separation*.

Probably the most widespread and effective

type of physical weathering is *frost shattering* which takes place when ice is formed in the joints and pore spaces of the rock. When water freezes in rocks it expands and sometimes attracts more water from other parts of the rock mass, which in turn freezes and expands. Thus, rock fragments are levered apart, often along joints and bedding planes, producing angular rock debris by block separation which may form spreads of boulders or screes in hilly districts. The effectiveness of frost shattering is believed to be greatest in areas where temperatures frequently fluctuate around freezing point, so that there are a large number of cycles of freeze-thaw. Also, water must be present, so that frost shattering is less effective in cold, arid areas.

Physical weathering also occurs as a result of temperature changes which take place well above freezing point. Such changes cause a rock mass to expand and contract, but the effect is restricted to a shallow layer near the surface because rocks are poor conductors of heat. This process may result in exfoliation. Igneous rocks in particular are composed of many different minerals which have different coefficients of expansion. Thus, temperature changes may create stresses within the rock mass, and granular disintegration may occur. Some doubt has been expressed about the effectiveness of this process since it is evident that little weathering has taken place in some desert areas where temperature changes are particularly great, and it is possible that the effectiveness of this kind of weathering is related in some way to the presence of water in the form of dew, fog or occasional rains.

Moisture is also necessary for the weathering of rocks in hot desert areas and on coastlines by the growth of salt crystals. In deserts, ground water may be drawn to the surface of porous rocks and evaporate. Tiny salt crystals may be formed and these may disintegrate the surface layer so that it either falls away as sand or is removed by exfoliation. This form of weathering may produce shallow caves, particularly near the base of a cliff.

Some of the rocks that now exist either at the earth's surface or at a shallow depth once existed at a much greater depth in the earth's crust. They have been exposed at the surface by the removal of the overlying rock. This release of pressure has allowed the rocks to expand, thus creating cracks (joints) roughly parallel to the ground surface, dipping into valleys and arching over hills. The process is known as unloading or sheeting by pressure release. Figure 1.8 shows pseudo-bedding in a mass of granite, which is believed to have originated in this way. The term 'pseudo-bedding' suggests that this type of jointing resembles stratification in sedimentary rocks. In some rocks, under certain conditions, pseudo-bedding may assist the occurrence of exfoliation.

Chemical weathering

Chemical weathering involves the chemical breakdown of the minerals of which rocks are composed, thus resulting in the disintegration of the original rocks. On a world scale it is much more effective than physical weathering. The chief agents of chemical weathering are oxygen, water, carbon dioxide and organic acids produced by the decay of vegetation. This kind of weathering reaches its greatest importance in hot, wet climates. Prominent jointing in rocks increases the surface area upon which chemical processes can operate. Of the main minerals which make up the igneous rocks, olivine, plagioclase-felspar and augite are more susceptible to chemical weathering than quartz, orthoclase-felspar and mica. Hence, basic igneous rocks such as basalt and gabbro are more likely to be strongly weathered than acid igneous rocks such as granite. Among the sedimentary rocks, limestone and chalk are particularly susceptible but sandstones are less so. Chemical weathering is particularly effective if it operates under a cover of soil. This is not only because various organic acids are produced within the soil but also because the soil cover ensures that the weathering surface is kept constantly wet. In general, chemical weathering produces regolith of much finer texture than physical weathering. Instead of screes and other types of coarse rock debris, one of the chief products of chemical weathering is clay.

The simplest type of chemical weathering is solution by water, but this is comparatively rare because so few minerals found in rocks are soluble in water. *Hydration*, however, involves the addition of water to a mineral. This may cause the mineral to expand and thus to create stresses within the rock in which it occurs. Thus, hydration may contribute towards physical weathering processes. *Hydrolysis*, on the other hand, is a chemical reaction between a mineral and water and is the most common chemical weathering process over the earth's surface. Basalt rock is changed into bauxite by hydrolysis, and this process is re-

A Dartmoor tor

Contrasting patterns of clints and grikes near Malham Cove

sponsible for rock decomposition to a great depth in warm, moist climates. *Oxidation* is also widespread over the world. In this case rock minerals react with oxygen to produce oxides or hydroxides. The waste mantle frequently becomes red or yellow in colour. Many of the red and yellow soils found in tropical areas are the result of this process. *Carbonation* is the weathering process performed by a weak solution of carbon dioxide in water. Carbon dioxide is usually abundant in soils, so soil water becomes a weak solution of carbonic acid which is capable of weathering carbonate rocks such as chalk and limestone. For this reason it is believed that the weathering of limestone is much more effective under a cover of soil than on outcrops of bare rock.

Figure 1.8(*a*) illustrates the evolution of a limestone pavement where the dominant weathering process is carbonation. In such a limestone area weathering concentrates its attack upon the joints and bedding planes but the weathering process has clearly been less effective at location X, where the grikes are comparatively narrow and the clints broad and tabular, than at Y where the clints are much narrower and the grikes wider. It is suggested that this variation is due to differences in the length of time that the limestone surface has been covered by drift or soil. X has been exposed to the atmosphere for a longer time than Y, so chemical weathering is less intense at point X. In the part of the pavement still covered by drift the grikes are very wide and the clints are narrow.

Figure 1.8(*b*) illustrates the most widely accepted theory of the formation of a granite tor. Such tors rise to about 10 m above the general level of a granite plateau such as Dartmoor. They are believed to have been formed by deep weathering in a warm, moist climate (probably in Tertiary times) along sets of radial joints and pseudo-bedding which is probably the result of pressure release (page 8). Where the joints are further apart the bedrock remains in place, and the blocks are separated by weathered clefts (as at X), but where the joints are more closely spaced (as at Y) the weathered blocks have been loosened and sometimes have fallen from the tor.

MASS WASTING

Once a sheet of regolith has been created on the land surface by the various weathering processes it is ready for movement downslope. In humid areas it will eventually reach river courses and then be carried to the sea. The various processes by which regolith is transported downslope towards river courses are referred to as mass wasting. Regolith tends to move down hillside or valleyside slopes at varying speeds and in different ways. The general principles are illustrated in Figure 1.9. This diagram shows a stationary mass of regolith on a slope. It remains stationary because the shear force generated by its weight is insufficient to overcome the friction at its base that holds it in place. If, however, the slope were to be steepened, the shear force would increase and would eventually overcome the friction. Also, if the plane of contact between the regolith and the underlying rock were to be lubricated in some way, reduced friction might allow the regolith to move downslope. There is also the possibility that the internal coherence of the regolith might be weak. In this case, although the base of the regolith might stay in place, its upper layers might flow downslope.

Types of mass movement on slopes

Types of mass movement on slopes vary to a great extent according to the water content of the moving mass of debris which influences its plas-

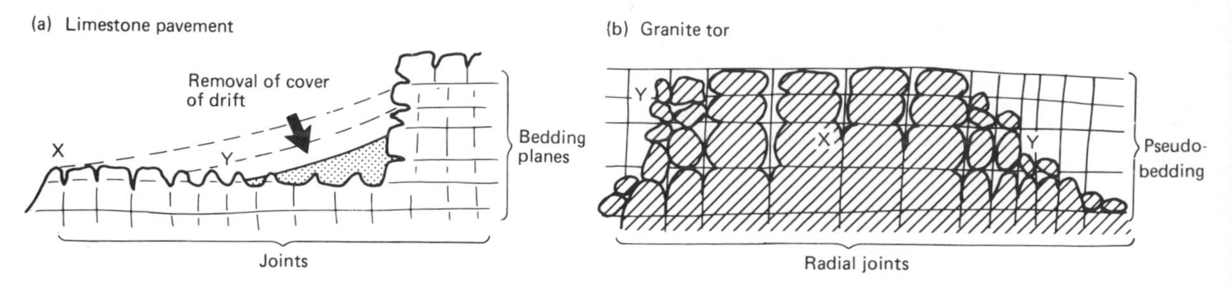

Fig. 1.8 Weathering of limestone and granite

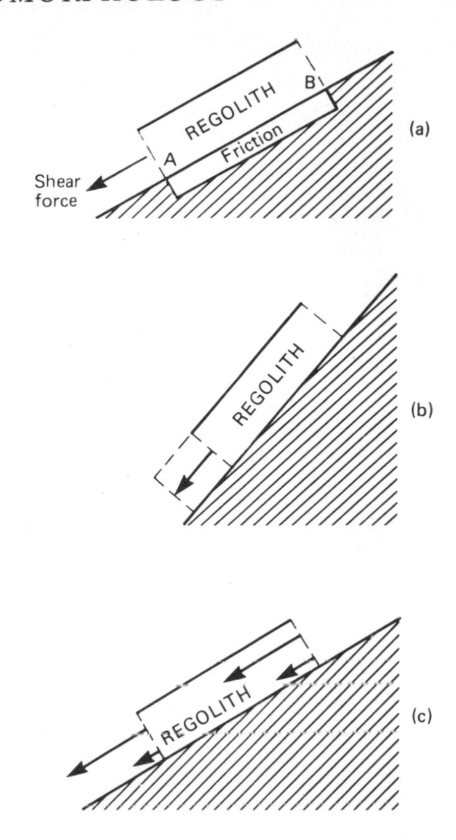

Fig. 1.9 The movement of regolith on slopes

ticity and the amount of friction that exists at its base. A common type of mass movement that is the result of heavy rainstorms is surfacc wash. In this process the impact of raindrops may loosen particles from the surface of the soil and these may be washed downslope. The result may be either a general movement of particles in the form of a sheet over the whole surface (*sheetwash*) or a concentration of run-off into small channels (*rillwash*) which may eventually develop into gullies. This kind of mass movement commonly occurs in semi-arid areas where there is only a sparse vegetation cover to protect the soil from heavy rainstorms. In humid temperate climates the soil is usually protected by trees with a litter of fallen leaves or grass which forms a continuous turf with a dense mat of roots.

In *mudflow* (Fig. 1.10(*a*)) a greater depth of soil is affected and this type of mass movement has a close resemblance to stream flow. If the subsoil is relatively impermeable, heavy rain may cause the water table to rise above the level of the base of the soil, thus reducing the friction between the soil and the subsoil. Also, the wetness of the soil itself provides efficient internal lubrication so that it is able to flow freely. In mudflow large masses of mud and water move down hollows in the slope and then expand into a delta shape as the gradient becomes less steep.

Soilflow (Fig. 1.10(*b*)) is a slower type of movement. In this case the moving mass of soil has a lower water content. Lubrication of the basal shear plane is not so efficient as in mudflow. To some extent therefore downslope flow is held back by friction at the shear plane. This process may leave lobes of soil that are convex on the downslope side. A special type of soilflow known as 'solifluction' occurs in areas that have an annual cycle of freeze and thaw. During frost the soil is heaved upwards. When the thaw comes the soil subsides and the release of meltwater lubricates the soil mass so that it is able to flow downslope.

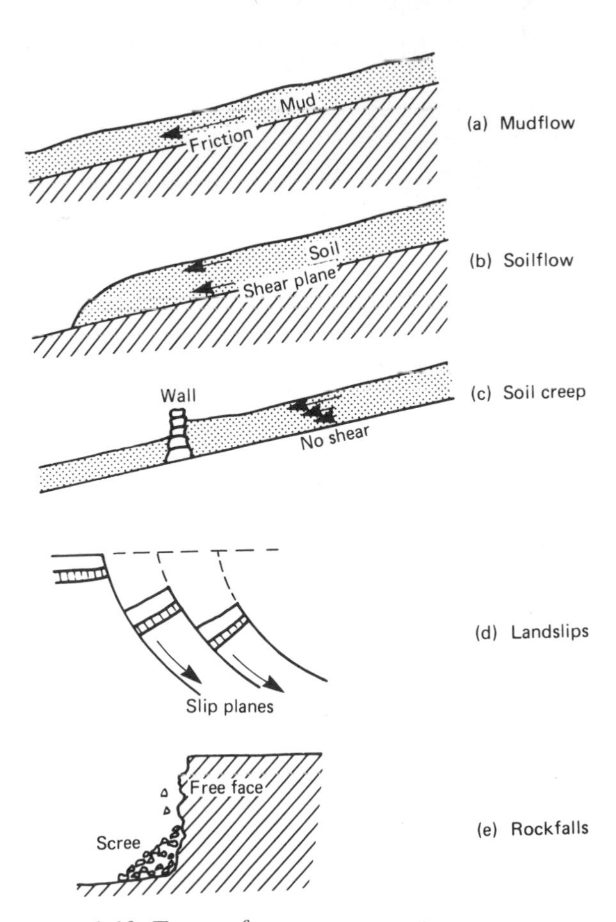

Fig. 1.10 Types of mass movement

A landslip. Note the rotation of the strata

An even slower type of mass movement is *soil creep*. In this case little or no movement takes place at the base of the soil. Movement is very slow but in time soil can accumulate on the upper side of walls and hedges (Fig. 1.10(*c*)). This process can also cause trees to be tilted away from the vertical.

The main distinction between the types of mass movement that have been described above is that the water content of the moving mass of soil differs. A high water content results in mass movement so rapid that it can be watched in operation as in the case of mudflow. The other factor is the steepness of the slope upon which mass movement occurs. All these types of mass movement are more likely to occur on steep slopes than on gentle slopes.

Other types of mass movement owe nothing to the lubrication of the regolith or soil by the presence of water. In *landslips* (Fig. 1.10(*d*)) there is no flow and no internal lubrication of the slipping rock mass. In some cases, however, the landslip may be partly the result of lubrication of the slip plane by percolating water. On the coasts of Kent and the Isle of Wight large blocks of chalk have slipped seawards over a bed of clay lubricated by percolating water. *Rockfalls* (Fig. 1.10(*e*)) owe nothing at all to the presence of water, except perhaps in the preliminary weathering phase. A rockfall is simply a fall of rock material that has been dislodged. The fallen fragments accumulate to form a scree at the base of the free face.

SLOPE PROCESSES

Various hypotheses have been formulated concerning the evolution of hillside slopes. These may be grouped into two major classes: slope decline (Fig. 1.11) and slope retreat (Fig. 1.12).

Slope decline

The principle of slope decline is associated with the 'cycle of erosion' or 'geographical cycle' suggested by W. M. Davis towards the end of the nineteenth century. This cycle begins with the uplift of an area to a greater altitude (stage 1 in Fig. 1.11(*a*)). At this stage the surface of the land would already have some relief because streams would have caused erosion while the uplift was taking place. These streams would continue to deepen their valleys, but for some time the highest ground would be unaffected. Hence, relief (the difference in altitude between summits and valley floors) would increase. Relief would reach a maximum between stages 2 and 3 in Figure 1.11(*a*). In stages 3 and 4 the deepening of the valleys would be much reduced and the emphasis would change to the wasting away of the inter-stream uplands by slope processes. While all this was taking place the steepness of valleyside slopes would be steadily reduced (Fig. 1.11(*b*)). Davis based his explanation upon the concept of a graded waste sheet, which is one whose downslope transporting ability is just equal to the task of transporting the regolith down the valley sides. In the early stages of the cycle perfect grading might not be achieved. Rock outcrops might form bare cliffs from which weathered fragments would fall as soon as they were loosened. Also, depressions might exist in the valleyside slopes which would have to fill with weathered debris before material could be transported across them. Such rock outcrops and depressions were regarded as similar to waterfalls and lakes which indicate a lack of grading in the long profile of a river. In the early stages of the cycle, therefore, graded waste slopes would have to be fairly steep so as to be able to transport coarse debris. With the passage of time, rock waste would become finer and finer and could therefore be transported down gentler slopes. Thus, with the passage of time, according to Davis, valleyside slopes would become less and less steep.

Slope retreat

Since the time of Davis attention has turned more towards the idea of slopes retreating rather than declining, although usually the retreat of a slope may be the cause of the slope's decline in steepness, as can be seen in Figure 1.12(*a*) which illustrates a suggestion of W. Penck. At first a steep rock cliff (*free face*) exists. When this is attacked by weathering processes all the rock waste falls to its base where it is assumed that it will be removed by a river. Thus, a complete slice of the free face can be removed apart from the lowest fragment which has no slope down which it can fall. Hence, the lowest fragment (marked X) remains in place. This process is then repeated, so that the free face moves backwards (parallel retreat) and is replaced by a gentler slope (the basal slope). In this way an appearance of slope decline is produced by the process of slope retreat. The diagram (Fig. 1.12(*a*)) gives the slightly wrong impression that the basal slope would be stepped. This is not so. It simply arises because the size of the basal fragments has been magnified in order to illustrate the process. A similar process is illustrated in Figure 1.12(*b*). Here, weathering of the free face results in the accumulation of scree which covers the lower part of the free face and reduces the effectiveness of weathering. Thus, as the upper part of the free face continues to retreat parallel to itself the lower part is protected from weathering by the accumulation

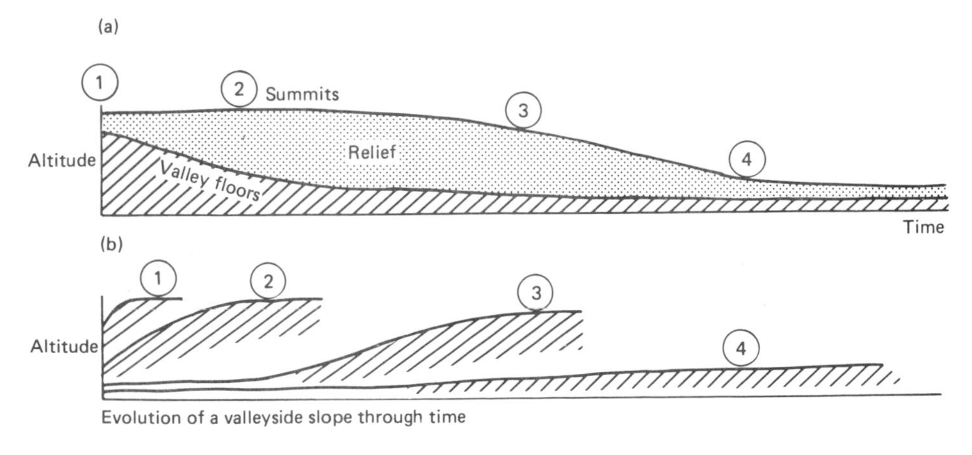

Fig. 1.11 Slope decline

of scree. Thus, the buried rock surface comes to have a convex profile reflecting the slow upward movement of the protection from weathering offered by the scree. Eventually, the free face disappears and the summit becomes smoothly convex. Whilst all this is taking place finer rock debris may be washed from the scree slope and may accumulate beyond it to form a more gently sloping wash slope, thus completing a concavo-convex hillside profile. The slope angle of the buried rock surface may vary in steepness. If weathering of the free face produces very coarse debris the scree will tend to grow upwards more rapidly, thus giving greater protection to the free face, which will be rapidly buried by scree. Thus, the buried convex rock surface will be steeper. Figure 1.12(*c*) illustrates a case in which both the free face and the scree slope (i.e. constant slope) retreat parallel to themselves. The constant slope here does not grow upwards and begin to cover the lower part of the free face. Instead, weathering processes reduce the rock fragments in the scree to finer particles and these are carried across the pediment by surface wash. In this example (Fig. 1.12(*c*)), the highest land shown (i.e. above the free face) is referred to as a waxing slope because, as the process of parallel retreat continues, it is replaced by a steeper slope (the free face). The pediment in Figure 1.12(*c*) is referred to as a waning slope because it becomes less steep as it develops. Processes such as these have been thought to have been important in the creation of the pediment-inselberg landscapes of much of tropical Africa which are dominated by large areas of almost level pediments interrupted here and there by abrupt changes of slope at the base of steep-sided inselbergs.

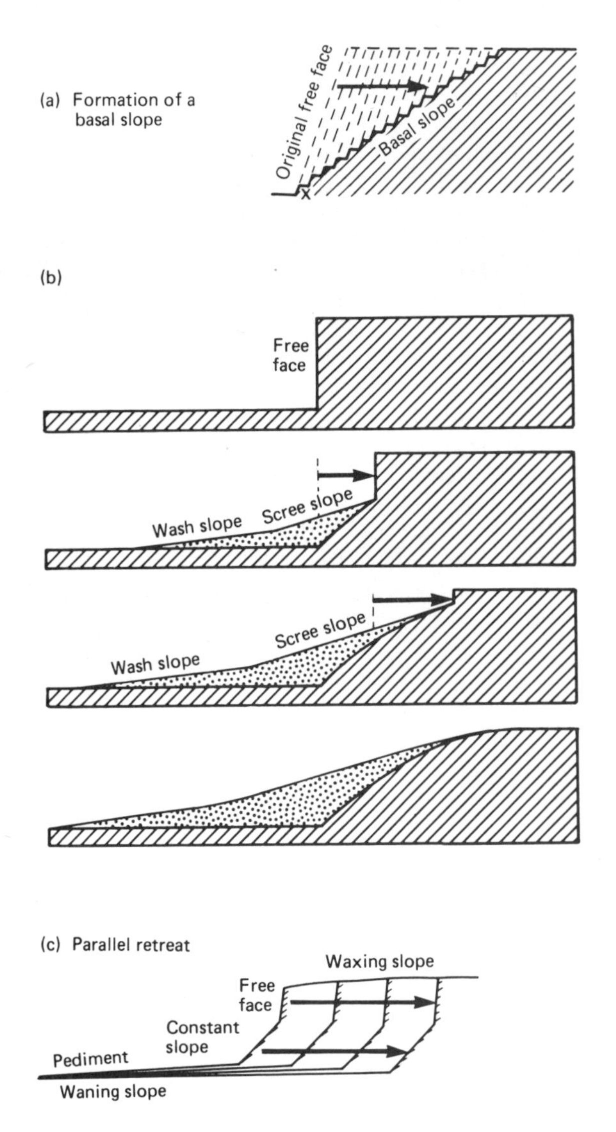

Fig. 1.12 Slope retreat

Other influences on slope form

The form of slopes can also be influenced by the characteristics of the types of rock upon which the slope is created. If the whole of a slope is made up of exactly the same type of rock it is unlikely that irregularities such as free faces will be created. The slope is more likely to take on a concavo-convex profile. Figure 1.13(*a*) shows the possible influence of a resistant cap rock forming the summit of a hill, underlain by a less resistant rock. The rocks could represent a weak sedimentary rock overlain by a sill composed of igneous rock. The resistant layer forms a free face on each side of the hill, beneath which is a constant slope down which rock waste is transported. If the rock waste could not be transported downslope it would build upwards and the free face would be covered by it. In this case parallel retreat takes place until the cap rock is completely removed. The slope would then assume a more convex profile and might become less steep than when the cap rock was present.

Hard limestones, such as those found in the Yorkshire Dales, tend to form convex or recti-linear (straight) slopes because their permeability reduces the effectiveness of rainwash. In detail,

A scree slope in the Similkameen Valley of British Columbia

A cuesta in the Brecon Beacons, Wales

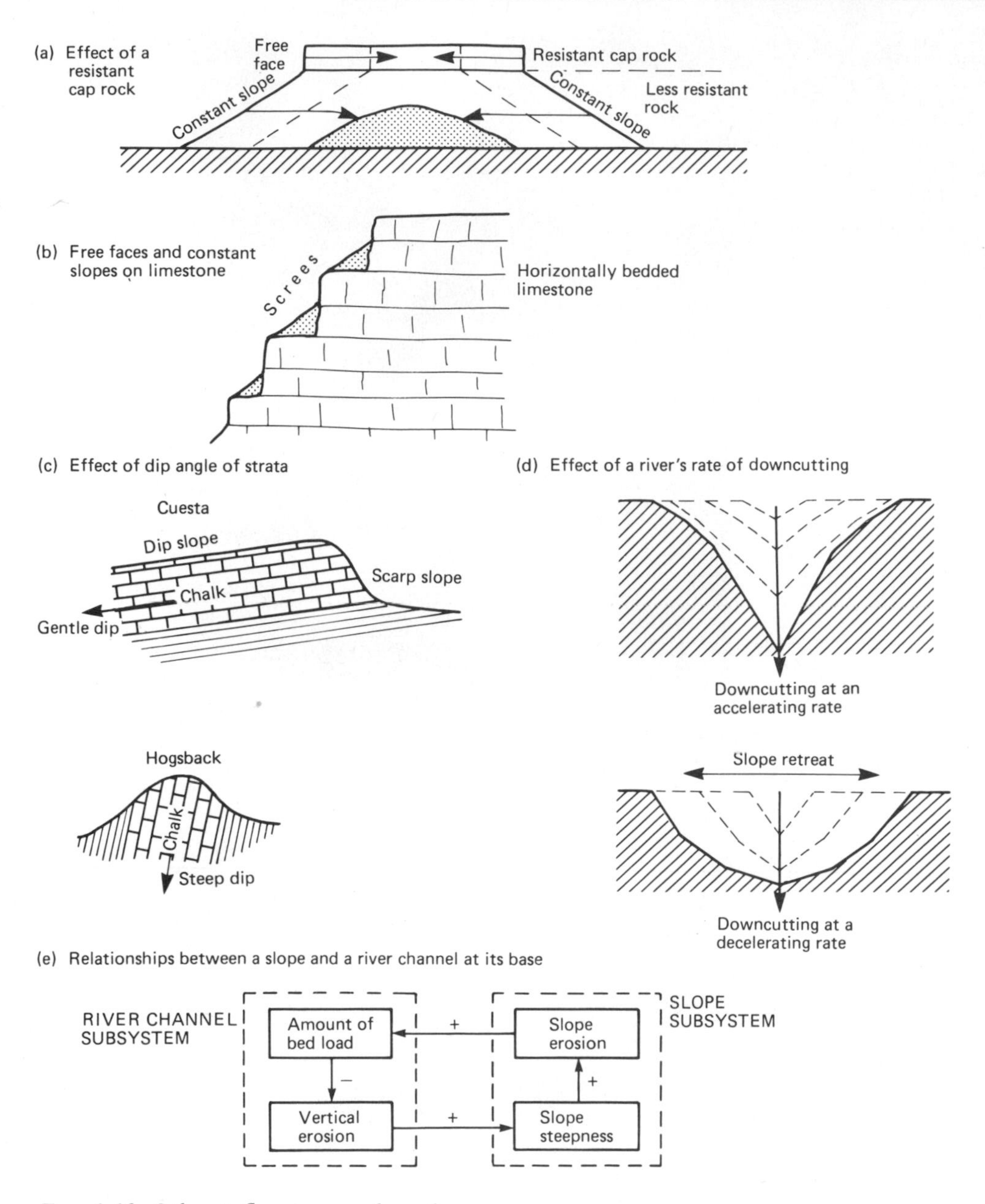

Fig. 1.13 Other influences on slope form

almost vertical free faces tend to be formed beneath which are constant slopes composed of scree that has weathered from the free faces (Fig. 1.13(*b*)).

Where a thick layer of chalk is associated with weaker rocks, such as clays, *cuestas* are commonly found, with scarp slopes and dip slopes (Fig. 1.13(*c*)). Here, the type of slope depends to a great extent on the angle of dip of the strata. Where the dip angle is relatively small, steep scarp slopes and gentle dip slopes are characteristic. Where the strata dip more steeply, the slope tends to become more symmetrical and a *hogsback*, with equally steep slopes on each side of the ridge, may be formed. Such a hogsback has usually been reduced to a much lower altitude than a cuesta.

According to Penck, the development of slopes may be controlled to a great extent by the rate at which the streams at their base are able to erode vertically. If the stream is downcutting at an accelerating rate, convex slopes tend to be formed (Fig. 1.13(*d*)). Concave slopes can result from the stream eroding vertically at a decelerating rate.

Hillside slopes and the streams at their bases are thought to be related to one another in an interlinked system of cause and effect relating to the slope itself and the river channel at its base. Such a system of cause and effect is illustrated in Figure 1.13(*e*). In this diagram the arrows and the plus and minus symbols indicate the following chain of events. An increase in vertical erosion by the stream will tend to increase (+) the steepness of the base of the slope. This increase will tend to increase (+) the bed load of the river (because mass wasting will be speeded up). The increase in bed load will then decrease (−) vertical erosion by the river. A system such as this is said to possess 'negative feedback'. An increase in one of the elements results in a certain amount of change through the whole system but this is limited by the characteristics of the system. In other words the system is self-regulating and after a change in any of the elements it regains a 'steady state'. Negative feedback which results in a return to a steady state can be recognised in a system by the existence of an odd number of negative relationships between the elements. The system shown in Figure 1.13(*e*) has only one negative relationship. Note that the pluses and minuses do not stand for increases and decreases. They relate to the direction of change as compared with the preceding element.

Slope forms also depend to some extent upon climate. In humid temperate climates, for example, vegetation protects the soil from surface wash. The soil is usually moist and this encourages soil creep, tending to produce slopes that are smoothly convex at the top, mainly rectilinear lower down and concave at the base. In deserts, on the other hand, slopes are either very steep free faces, screes or level rock pediments. Alternating cliffs and screes occur in succession up hillsides. In very dry climates little fine-grained regolith is produced. Sometimes there is a very sharp angle between the level pediment and the base of the rock slope. Slope retreat is fastest of all in hot, wet tropical climates where chemical weathering is very intense. Mass wasting takes place at a rapid rate, so slopes are mainly convex and there are few free faces.

1.3 Rivers and their basins

HYDROLOGY OF THE RIVER DRAINAGE BASIN

A drainage basin is the catchment area from which a river system obtains its supplies of water (Fig. 1.14(*a*)). Precipitation falls over the area bounded by the major watershed and water makes its way either over the ground surface or by underground routes to the various streams, which then converge to form a single trunk stream that carries water to the basin outlet. The drainage basin is separated from neighbouring basins by the major watershed. Minor watersheds separate the drainage basins of the tributary streams. Thus a large river drainage basin may contain a complex hierarchy of minor drainage basins.

The hydrological cycle in a drainage basin

Figure 1.14(*b*) is a pictorial representation of the hydrological processes that operate in a drainage basin. Figure 1.14(*c*) shows the same processes in the form of an abstract 'systems' diagram. Precipitation from the atmosphere first enters the vegetation subsystem by falling on trees, bushes or grass. Some of it passes quickly through the vegetation cover by falling through spaces in the canopy (*throughfall*), some rather less quickly by running down stems and branches (*stemflow*) and dripping from leaves (*drip*). Some evaporates directly from the vegetation. The moisture that passes through the vegetation subsystem next reaches the surface subsystem (the ground surface). Here, it may be stored in puddles, ponds or lakes and may be returned to the atmosphere by direct evaporation or by transpiration via the vegetation subsystem. It may also move by overland flow to stream channels or it may sink into the ground (*infiltration*) and enter the soil subsystem. In the soil, water may flow laterally through the soil (*throughflow*), especially if the ground surface has a considerable slope (Fig. 1.14(*b*)) and reach river channels, or it may, particularly in a dry spell,

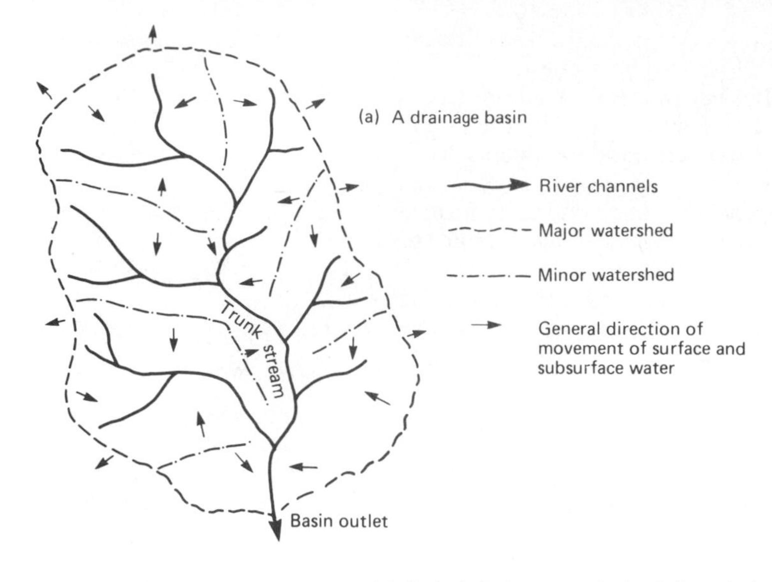

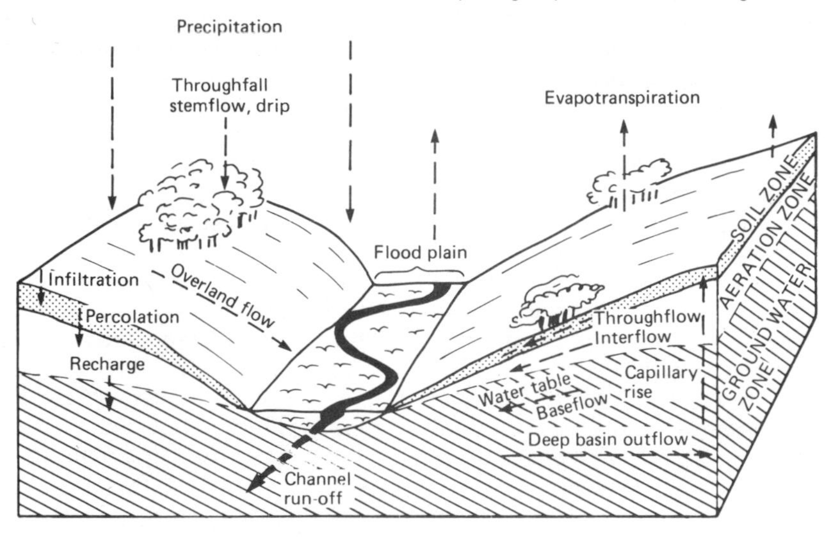

Fig. 1.14(a)(b) Hydrology of a drainage basin

return to the ground surface by capillarity. Water from the soil subsystem may also sink deeper and reach the aeration zone subsystem and even the ground-water subsystem that lies below the water table. From here it may move laterally by *interflow* and *baseflow* and reach river channels. Even from these deep regions, in certain conditions, water may rise back to the ground surface by capillarity. From the ground-water subsystem some water may flow at depth beyond the confines of the particular drainage basin (*deep basin outflow*). Water also leaves the drainage basin by channel run-off along the river channel itself. The river channels and flood plains receive water from direct precipitation and by flow from surface and sub-surface subsystems, and they also lose water by evapotranspiration.

The drainage basin as a whole can be described as an open, cascading system. It is an 'open' system because it receives inputs of energy (solar radi-

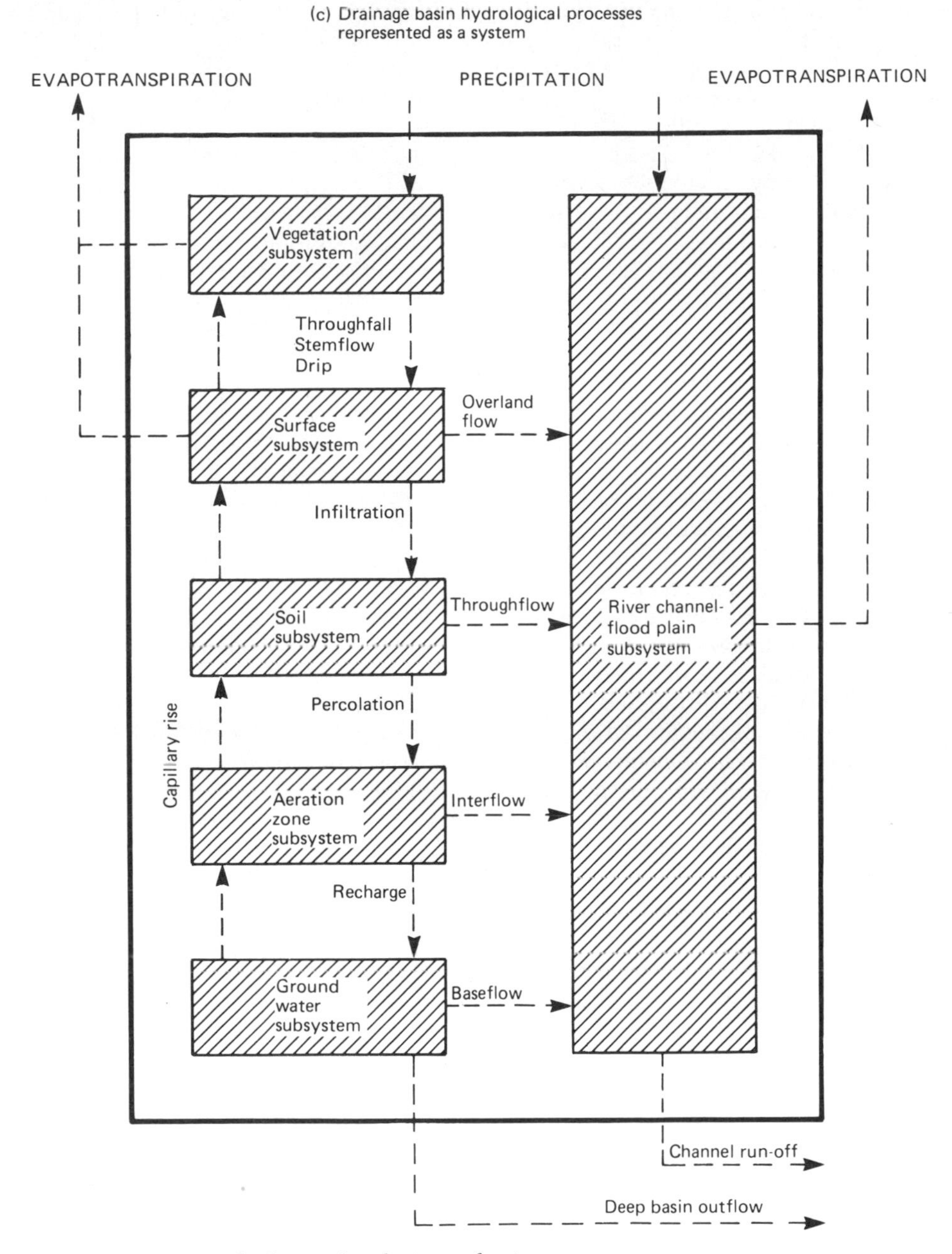

Fig. 1.14(c) Hydrology of a drainage basin

ation) and mass (precipitation) from outside the system and it delivers outputs of water and eroded debris to the outlet of the basin. It is described as a cascading system because the output from one subsystem becomes in turn the input of the succeeding subsystem. These exchanges are governed by 'regulators' within each subsystem which control the amount of water that is able to travel along any particular route in the system. Each regulator establishes a 'threshold' – a maximum flow of water along a route that cannot be exceeded. Any available water greater than this must move along an alternative route or must be stored.

Details of the operation of the subsystems
The amount of precipitation that passes through the vegetation subsystem to reach the ground surface is determined by the extent to which precipitation is intercepted by the vegetation and then evaporated (Fig. 1.14(*c*)). Much rain that falls on the vegetation may never reach the ground surface. But when the vegetation is fully saturated any additional precipitation must fall to the ground. The greatest loss of water through interception tends to occur at the beginning of a rain shower and to reduce steadily as the rain continues and the vegetation becomes wetter and wetter. Hence, the greatest loss of water from the system through interception and evaporation takes place when rain falls in sporadic light showers rather than as continuous rain. Interception by coniferous evergreen trees tends to be greater than that by broad-leaved deciduous trees. Water tends to cling to needle-shaped leaves but the droplets tend to run together and drip from broad deciduous leaves. Also, deciduous trees are bare of leaves in winter, so at this time little interception can occur. In summer, air can circulate more freely through coniferous woodland than through deciduous, and this can encourage quicker evaporation. Snow is more likely than rain to be intercepted by the vegetation but, on melting, it will rapidly slide off and fall to the ground. Grassland has a lower interception capacity than either coniferous or deciduous trees. An unusual characteristic of the hydrological cycle exists in the island of Tenerife (Canary Islands), where a belt of pine forest encircles the lower and middle slopes of Mt Teide (3717 m) and is quite frequently enveloped in cloud. In this semi-arid climate fog drip from the pine needles has been known to exceed the annual precipitation. An important regulator is related to the surface subsystem. This determines the proportion of water that evaporates or flows over the surface to stream channels or infiltrates into the soil subsystem. In general, excessive overland flow is undesirable because it will reach river channels very quickly and cause sudden floods and increased erosion in river channels. Hence, a high infiltration capacity is usually regarded as desirable. A high infiltration capacity is usually dependent on water being held in place on the ground surface and being given the time to infiltrate. Thus, a rough ground surface and generally level ground encourage infiltration. Smooth, steep slopes encourage overland flow. Vegetation generally favours infiltration because grass provides a rough surface and forest provides a leaf litter and possible infiltration channels via the roots of the trees. Sandy soils encourage

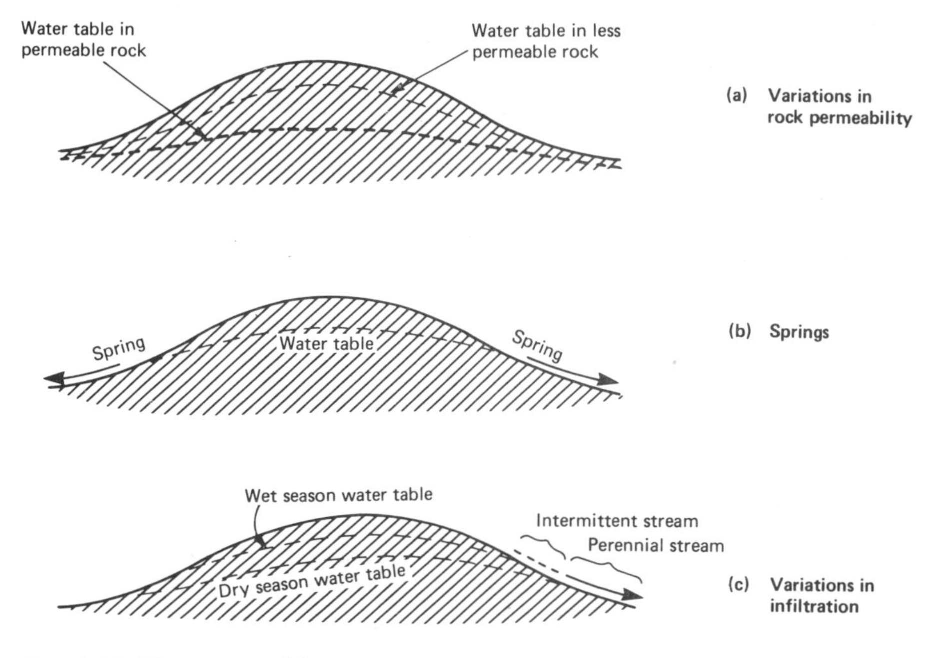

Fig. 1.15 The water table

infiltration; clay soils soon become saturated. Very heavy rain showers may soon cause the infiltration capacity of the ground surface to be exceeded, so that overland flow begins. A steady drizzle gives the most efficient infiltration. A cover of snow slowly melting may deliver water to the soil surface at a rate which allows steady infiltration, but a rapid thaw could lead to overland flow. The water table separates the aeration zone subsystem from the ground-water subsystem. Beneath the water table the rock is saturated with water. In a very porous or well-jointed rock the water table will tend to be horizontal because the water will encounter little resistance to movement through the rock (Fig. 1.15(*a*)). In a rock of finer texture, however, movement of ground water is hindered and the water table tends to reflect the shape of the ground surface, being domed upwards under hills. If the water table reaches the ground surface on a hillside it may form a spring (a small river channel) (Fig. 1.15(*b*)). The level of the water table tends to fluctuate according to the amount of water that penetrates down to the ground-water subsystem, which depends upon the balance between precipitation, overland flow and evapotranspiration at the ground surface. This can result in the creation of intermittent streams (Fig. 1.15(*c*)).

Channel run-off

Some of the precipitation that falls on the drainage basin eventually makes its way to river channels and becomes channel run-off (Fig. 1.14(*c*)) or, in other words, stream discharge. It is useful here to make a distinction between quickflow and ground-water flow. In *quickflow* there is only a short interval between the time when precipitation first reaches the drainage basin and the time when it is delivered to the river channel. Quickflow is usually taken to include only overland flow and soil throughflow (Fig. 1.14(*c*)). It is the short-term response to variations in the intensity of precipitation. Since quickflow reflects precipitation so closely it can be a major cause of the occurrence of floods. *Ground-water flow* on the other hand takes a deeper route from the ground surface to the river channel and travels much more slowly. It is usually taken to include interflow and baseflow (Fig. 1.14 (*c*)). Ground-water flow only arrives at the river channel a considerable time after the original precipitation reaches the ground surface. Thus, although precipitation occurs only intermittently, ground-water flow tends to arrive continuously at river channels. It tends to maintain a steady flow even through periods of drought, and sudden, heavy rainstorms have little effect.

Thus the relationship between quickflow and ground-water flow determines the character of the discharge of stream channels. Discharge fluctuates greatly in streams in which the quickflow element is great. Streams supplied mainly by ground-water flow maintain a more constant discharge. In a drainage basin, therefore, there may be three types of streams (Fig. 1.16). Perennial streams are supported by a fairly constant level of ground-water flow from below the water table and hence their discharge is the most constant. Intermittent streams, usually at a higher elevation, are only supplied by ground-water flow at times when the water table rises. Hence, they will only maintain a constant discharge at these times. At higher elevations still, ephemeral streams only flow when they are supplied by quickflow for relatively short periods immediately after precipitation has occurred. This sequence tends to occur in a drainage basin as one moves from the trunk stream along a tributary towards the major watershed.

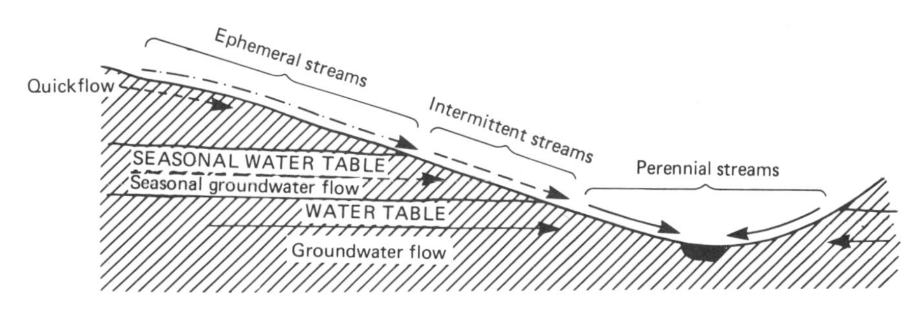

Fig. 1.16 Different types of streams in a drainage basin

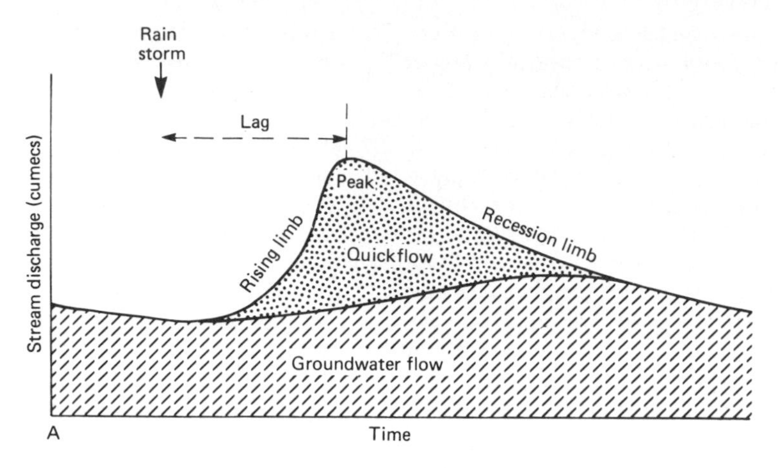

Fig. 1.17 Variations in stream discharge

The hydrograph of stream discharge

A hydrograph is a record of the flow of water along a stream channel measured at a particular point. Figure 1.17 illustrates some of the variations in stream discharge that may result from a shower of rain in a drainage basin upstream of the recording site. Stream discharge is measured in cumecs (cubic metres per second).

At the beginning of the time period (A in Figure 1.17) the stream's discharge consists entirely of ground-water flow and this has been gradually decreasing since the last rain period. When the rain storm occurs there is a time-lag before its full effect upon stream discharge is felt. During this period the increased discharge from the headwater streams is making its way to the recording point. Most of the increased discharge is contributed by quickflow but there is a steady increase in ground-water flow which continues after the time of peak discharge. Discharge as a whole decreases along the recession limb, but rather more slowly than the increase on the rising limb. Eventually, some time after the rain storm, the volume of ground-water flow begins to decrease. In a perennial stream, of course, the hydrograph always shows some discharge, which is provided by ground-water flow.

Physical factors influencing the shape of the hydrograph

The shape of a hydrograph recorded at the outlet of a drainage basin is influenced by the type of precipitation that falls on the drainage basin and the physical characteristics of the drainage basin. These factors have a strong influence on the relative proportions of quickflow and ground-water flow (Fig. 1.17) that reach the gauging station. Intense rainfall tends to produce a high proportion of quickflow since the soil may be incapable of accepting and storing such large amounts of water. Also, a long period of continuous rainfall can cause the water table to rise to the surface in many parts of the basin and thus produce a sudden increase in overland flow that will give the hydrograph a high peak. If rainfall is concentrated into a number of consecutive heavy showers these may produce a series of additional peaks on the recession limb. The melting of snow, combined with rainfall, can produce very high peak discharges especially if the ground is still frozen. In catchments with a semi-arid climate, rain may fall only occasionally, but with great intensity, and streams may only flow for a short period after rain has occurred. In such cases flash floods can occur and the hydrograph records only quickflow, with no ground-water flow at all. In a flash flood the rising limb of the hydrograph may be nearly vertical. Large drainage basins tend to produce higher peak discharges than smaller basins. On the other hand, smaller basins often occur in upland areas and have steeper slopes. Hence a small drainage basin may deliver water to its outlet very rapidly and thus produce a hydrograph with a very steep rising limb and a short lag time (Fig. 1.17).

The geology of the drainage basin has a great effect upon stream discharge. Basins with a chalk subsoil usually have a great capacity for the storage of water, and rainfall infiltrates readily since the chalk is porous. This tends to produce a fairly flat discharge curve dominated by a high proportion of ground-water flow. In chalk areas the quickflow element of stream discharge tends to increase gradually during a period of rain as the water table slowly rises and fingertip tributaries gradually extend into dry valleys (Fig. 1.16). In hard limestones, such as Carboniferous limestone, the subsoil is permeable but not porous. Channel flow can take place underground through widened fissures in the rock. Although underground lakes may regulate the flow of water to some extent, in these areas very high peak discharges may occur. If the rocks of a drainage basin are impermeable, such as clay, both limbs of the hydrograph may be fairly steep, but, of course, much would depend on other factors such as the steepness of slopes.

A high density of stream channels in a drainage basin tends to increase peak discharge. If there are many closely spaced stream channels relatively slowly moving water will reach them more quickly and become quickflow. The existence of natural storage facilities in the basin, such as lakes, swamps and flood plains can flatten the discharge curve. Channel run-off will be temporarily stored. Forested catchments tend to have a relatively low peak discharge and a greater amount of ground-water flow because forest vegetation tends to encourage infiltration.

The pattern of stream channels within the drainage basin can influence discharge. A high peak discharge can occur if most of the tributaries converge on the basin outlet. In this case, if a rain storm affects the whole basin, flood water from all the tributaries may arrive simultaneously at the basin outlet (Fig. 1.18(*a*)). If the drainage basin is long and narrow and a storm moves along it from the outlet (Fig. 1.18(*b*)), the peak discharge tends to be lower since each tributary contributes flood water in turn. It is possible for there to be a succession of peak discharges or possibly a 'plateau', with a fairly high discharge being maintained for a considerable time. If, in a narrow drainage basin, a storm moved towards the outlet,

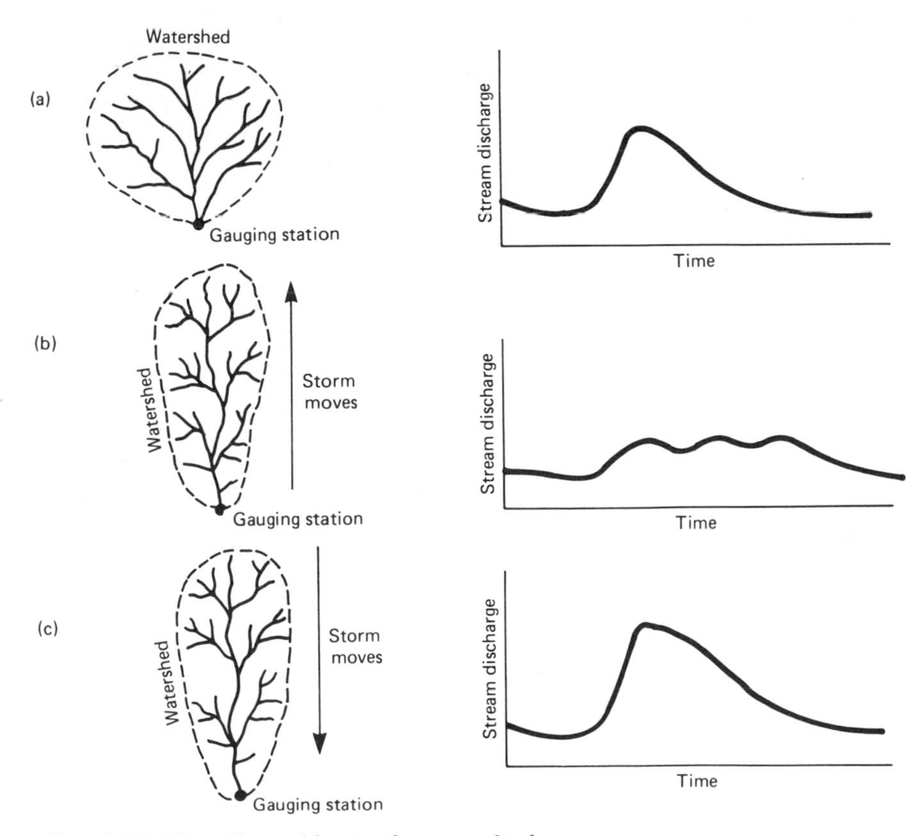

Fig. 1.18 The effect of basin shape on discharge

a very high peak discharge could occur since flood water from all the tributaries might arrive almost simultaneously at the gauging station (Fig. 1.18(*c*)).

Human factors influencing the shape of the hydrograph

Changes in rural land use can influence the hydrological system in a drainage basin. A rather extreme case, the removal of vegetation, is illustrated in Figure 1.19. This diagram shows that as vegetation is progressively removed the infiltration capacity of the surface subsystem is reduced (a negative relationship). This causes overland flow to increase (another negative relationship), which in turn increases erosion of the ground surface (a positive relationship), and this in turn causes a reduction in the infiltration capacity (a negative relationship). This will then cause further increases in overland flow and surface erosion followed by further decreases in infiltration capacity, and so on. The system illustrated in Figure 1.19 is therefore said to possess positive feedback. Positive feedback has an even number of negative relationships in the feedback loop (or of course no negative relationships at all).

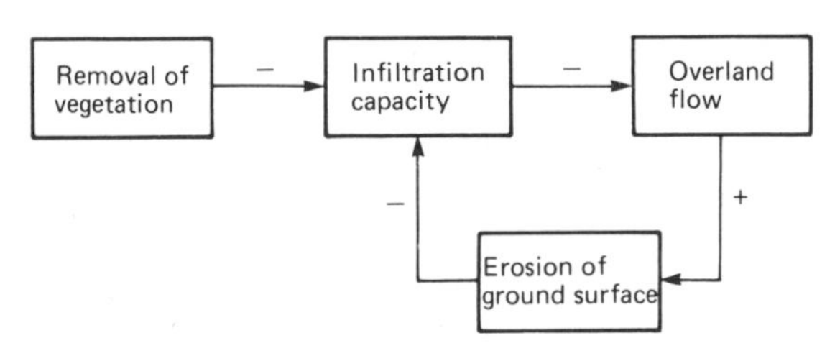

Fig. 1.19 Positive feedback resulting from the removal of vegetation

Changes in the type of vegetation or crop can also influence the hydrological system in a drainage basin. In general, coniferous evergreen woodland tends to give the most uniform stream discharge with the smallest amount of quickflow. If this is replaced by deciduous trees, quickflow tends to increase and there is a greater variation in stream discharge. If woodland is replaced by grassland the infiltration capacity of the ground surface is further reduced, partly because of the disappearance of the trees' root channels and also because the intensity of rainfall reaching the ground surface increases as a result of the reduction in interception. With a change from grassland to arable farming there tends again to be a reduction in infiltration. Ploughing may temporarily decrease overland flow and increase infiltration by detaining water in the furrows, but the soil surface tends to become compacted by rain beat. The installation of either open drains or underground tile drains in agricultural land clearly tends to increase quickflow. Also the draining of marshland is intended to accelerate channel run-off and thus tends to lead to an increase in peak discharge. The dredging of rivers may to some extent increase channel storage of water, but it also tends to accelerate channel run-off. The building of dams across rivers for water supply, power or irrigation gives man some direct control over stream discharge. Reservoirs, with some limitations, can smooth out variations in stream discharge by encouraging infiltration and percolation to the ground-water subsystem (Fig. 1.14(*c*)) and also by increasing evaporation. Urbanization of part of a drainage basin has a great effect upon its hydrology. Many surfaces in towns, such as roads, pavements and roofs, are virtually impermeable and drainage systems are designed to conduct water away from the town as rapidly and efficiently as possible. Hence, urbanization greatly reduces infiltration and consequently reduces the ground-water flow element of stream discharge. Within the town, flooding is regarded as undesirable, so natural streams are straightened and made more efficient for transporting water away from the urban area. Some parts of a town may be built on a flood plain. In this case embankments may be built to prevent flooding. This, of course, prevents the flood plain from acting as a natural storage area for flood water and also may improve the efficiency of the river channel in transporting water further downstream. Thus, quickflow from the drainage basin can move rapidly downstream. Hence, the hydrograph is affected in two ways. The peak discharge tends to be very great compared with the normal ground-water flow, and the lag time between the occurrence of a storm and the peak discharge is very small.

Seasonal variations in river discharge over the world

The most important influence on patterns of river discharge considered on a world scale is climate. The volume of river discharge reflects the balance between precipitation and evapotranspiration, and evapotranspiration is closely related to temperature.

In the tropics, generally temperatures are high throughout the year so evapotranspiration is always at a high level. Seasonal variations in river discharge therefore depend mainly on the seasonal distribution of precipitation. Along the equator, all-season precipitation usually provides a surplus over evapotranspiration. Hence, most rivers near the equator have a fairly constant discharge through the year, sometimes with maxima twice a year in periods of rather heavier rainfall. Northwards and southwards from the equator a tendency appears for rainfall to be concentrated in the hotter season. This tendency is particularly evident in the 'monsoon lands' of south-east Asia. Here, river floods tend to occur in summer and discharge shows a marked decrease in the dry winter. The Nile is an interesting example. It has its headwaters in Ethiopia (rainfall maximum in summer) and East Africa (all season rainfall). Hence, its maximum discharge is in summer. The Nile then flows northwards, crossing the Sahara Desert, and supplies a maximum discharge in late summer to Egypt, where very little rainfall occurs.

Outside the tropics the situation is more complex. In temperate regions there are not only seasonal variations in precipitation but also large seasonal variations in temperature which cause large variations in evapotranspiration. Also, many areas receive heavy falls of snow in winter which, because of the low temperatures, is mostly stored until spring. In the British Isles river discharge is usually slightly greater in winter than in summer. Precipitation is fairly constant through the year, but evapotranspiration is considerably greater in the summer season. In the coastlands of the Mediterranean Sea, maximum discharge occurs in autumn and winter, the time of maximum rainfall. In summer, many of the rivers dry up completely and river channels contain only pebbles and sand. This is because very little rain occurs and evapotranspiration is at a high level because of the very high temperatures. Rivers such as the Rhône can only maintain their flow through the summer because they are supplied with water from elsewhere. The Rhône is a particularly

A stream channel at the beginning of the dry season on the Mediterranean island of Rhodes

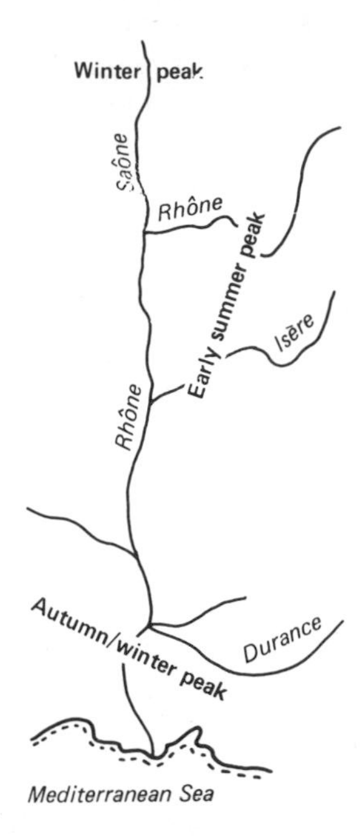

Fig. 1.20 Discharge in the Rhône Basin

interesting example. It rises in the Alps, where its maximum discharge occurs in early summer, from the melting of ice and snow in the mountains. Its northern tributary, the Saône (Fig. 1.20), has a maximum discharge in winter, the result of the all-season rainfall in central France with relatively high rates of evapotranspiration in summer. Further south, the Isère brings to the Rhône another early summer meltwater peak discharge from the Alps. Finally, nearer the Mediterranean coast, an autumn/winter peak occurs. All these peak discharges are superimposed on one another near the mouth of the Rhône to produce a complex succession through the year.

Many of the rivers of central and eastern Asia, and much of North America have two separate peak discharges each year. One of these occurs in early summer with the melting of the winter snows. The other occurs in August or September and reflects the summer rainfall maximum.

QUANTITATIVE CHARACTERISTICS OF DRAINAGE BASINS

Detailed studies have shown that stream channels and river drainage basins do not form random patterns. A considerable number of statistical regularities have been observed, and some of these have even been described as 'laws'. Quantitative analysis of the characteristics of drainage basins has allowed comparisons to be made between drainage basins in different climates and on different kinds of rocks.

General characteristics of a drainage basin

The general shape of a drainage basin may be described by using various descriptive terms such as 'circular' and 'triangular', but it is also possible to describe a basin's shape quantitatively, thus making it easier to compare the shapes of a number of basins. The 'form factor' describes a basin's shape by comparing the area of the drainage basin with the area of a square whose sides are equal in length to the length of the drainage basin from its outlet to the most distant point on the basin's boundary (watershed) (Fig. 1.21(*a*)). A compact drainage basin will have a form factor of relatively high value, approaching unity. Another measure of the shape of a drainage basin is the 'coefficient of compactness'. This compares the basin's shape with that of a circle which is the most compact plane shape in the sense that for a given area it has the shortest possible circumference. The compactness coefficient is calculated by dividing the length of the basin's perimeter by the length of the perimeter of a circle whose area is equal to that of the basin. A low value (near to unity) for this coefficient indicates a compact drainage basin.

Another general characteristic of a drainage basin is its 'drainage density'. This is calculated by dividing the total length of all the drainage channels in the basin by the area of the basin. It is expressed as km per sq. km (Fig. 1.21(*b*)). Drainage density tends to be low in areas of resistant rocks such as granite and some sandstones. In such areas, erosion of a channel can only be performed by quite large streams. Low drainage densities are also common in areas of permeable rocks. In chalk areas of southern England, for example, stream channels are very widely spaced and very low drainage densities are common. Clays and shales, which are both impermeable and weak, tend to produce high drainage densities, but these could be reduced by the occurrence of forest or grass surface cover which could increase infiltration and reduce overland flow. Weak clays in semi-arid climates where heavy rain storms occur occasionally may be carved by overland flow into the 'badland' type of

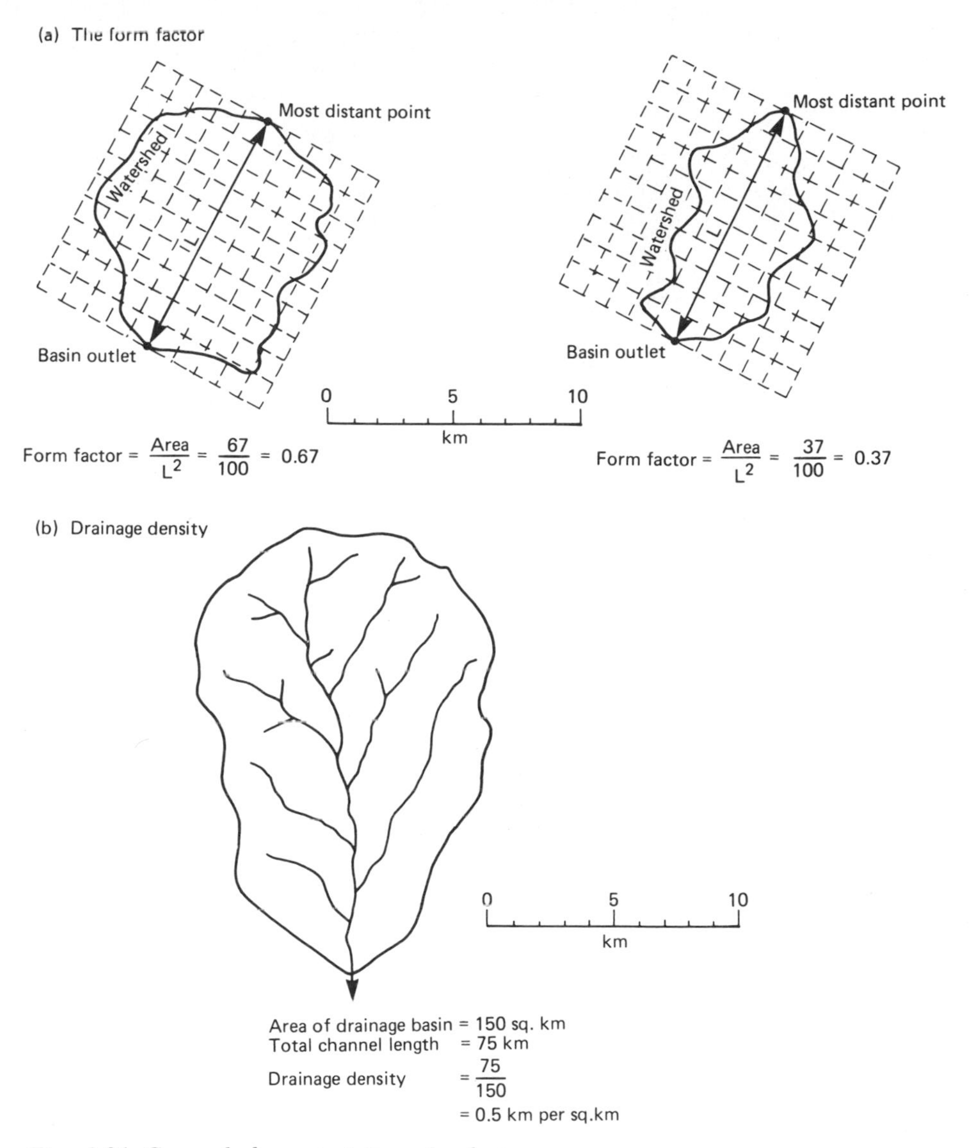

Fig. 1.21 General characteristics of a drainage pattern

landscape which may have a drainage density of up to 500 km per sq. km.

Stream orders

Methods have been devised for describing and comparing the networks of stream channels that exist in drainage basins. The basis of these methods has been the classification of segments of stream channels into a system of 'orders'. Figure 1.22 shows an idealized drainage basin in which the stream channels have been ordered by using a system devised by Strahler. In this system, the fingertip channels with no tributaries are classified as first order streams. If two first order streams join they form a second order stream, which may then receive first order tributaries with no change in its order. If two second order streams join, a third order stream is created. A third order stream may receive first and second order tributaries without changing its order. Higher order streams are identified in a similar way.

Other methods of ordering stream segments

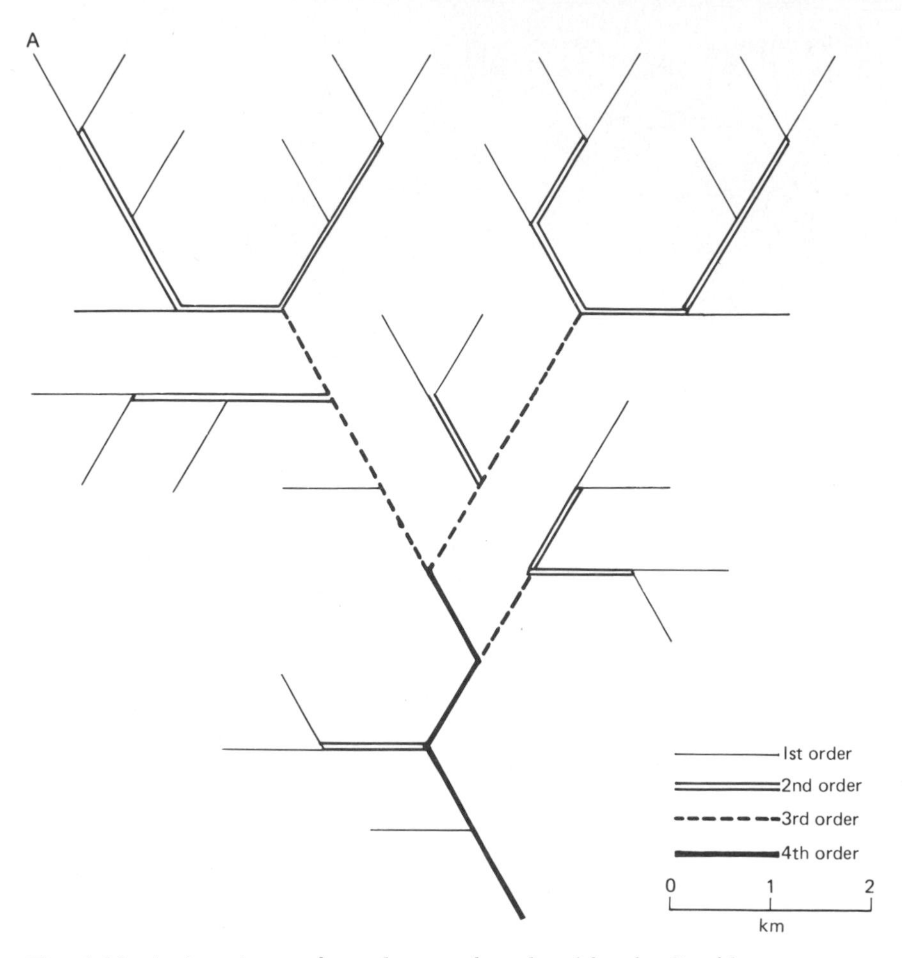

Fig. 1.22 An imaginary channel network ordered by the Strahler system

have been suggested by Horton and Shreve. Horton's system was the first to be devised. It is similar to that shown in Figure 1.22 except that the higher order stream segments are extended headwards either to the source of the longest tributary or along the straightest continuation of the channel direction at a confluence. By this principle the fourth order stream shown in Figure 1.22 would be regarded as extending upstream as far as the source labelled A instead of only as far as the confluence of the two third order streams. Shreve referred to orders as 'magnitudes'. The magnitude of a stream segment is the number of fingertip tributaries that supply it with water. The fourth order segment shown in Figure 1.22 reaches a magnitude of 27. In Shreve's system, therefore, the magnitude changes at every confluence.

Quantitative regularities in stream networks

Studies of stream networks using the stream ordering systems described above have revealed certain regularities in the relationships between the various orders of streams, particularly when the surface of the whole drainage basin consists of a single rock type.

One of these regularities relates to the number of stream segments in the various stream orders. The table below refers to the channel network shown in Figure 1.22.

Stream order	Number of segments	Bifurcation ratio
1	27	
		3
2	9	
		3
3	3	
		3
4	1	

From this table it is clear that each successively lower order has three times the number of stream segments possessed by the next higher order. The bifurcation ratio is 3. The bifurcation ratio is calculated by dividing the number of stream segments in any stream order by the number in the next higher order. This relationship is referred to as a geometrical progression and it is represented by a straight line when plotted on semi-logarithmic graph paper (Fig. 1.23(*a*)). Bifurcation ratios in general have values between 3 and 5. Higher values tend to occur in long, narrow basins where a trunk stream may be fed by many low order tributaries. The lowest possible bifurcation ratio is 2 since, according to Strahler's stream ordering system, two lower order streams are needed to produce a stream of the next higher order. A bifurcation ratio of 2 would also mean that the higher order stream would receive no extra tributaries in addition to the two lower order streams that created it.

A similar regularity tends to exist in respect of the lengths of the various stream segments of the different orders. High order stream segments tend to be longer than low order ones. The table below refers to the channel network shown in Figure 1.22.

Stream order	Mean length of segments	Length ratio	Cumulative mean length
1	1		1
		1.8	
2	1.8		2.8
		1.3	
3	2.3		5.1
		1.7	
4	4.0		9.1

In this case, the length ratio like the bifurcation ratio (above), tends to remain constant. If the cumulative mean lengths of stream segments are plotted against stream order on semi-logarithmic graph paper (Fig. 1.23(*b*)) a straight line tends to be produced.

A similar relationship to that shown in Figure 1.23(*b*) is also evident in respect of the sizes of the drainage basins of the streams of different orders within the large basin of the highest orders stream. Beginning with the first order drainage basins, the areas of successively higher order drainage basins increase in such a way as to produce a straight line rising towards the right on semi-logarithmic graph paper.

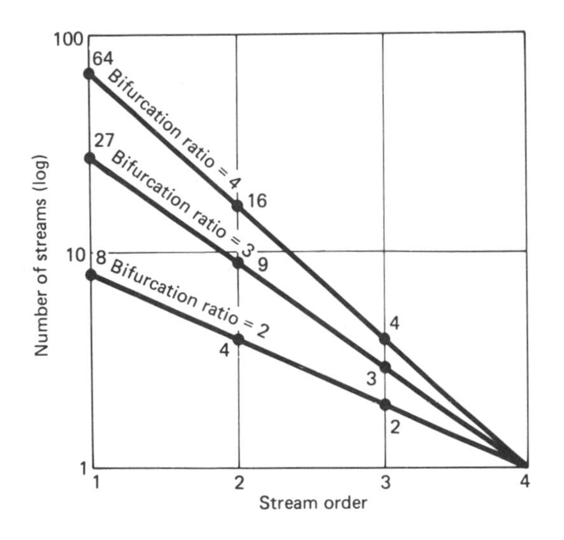

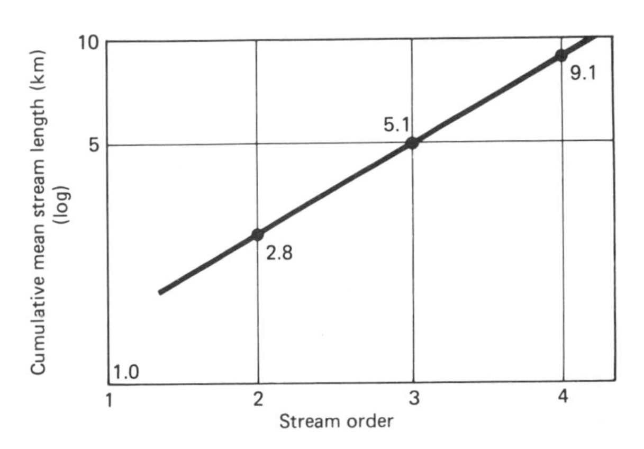

Fig. 1.23 Graphs showing regularities in stream networks

DRAINAGE PATTERNS

Stream networks can also be classified according to the nature of the geometrical patterns which they form on the surface of the ground. These patterns are related in the first place to the mode of origin of the stream system, but they also reflect the topographical and geological characteristics of the land surface upon which the network has evolved. The classification applies both to the streams themselves and also to the networks of which they form a part.

A dendritic drainage pattern

A dendritic drainage pattern has the appearance of the trunk, branches and twigs of a tree (Fig. 1.24(*a*)). Tributaries join each other and then join the trunk stream at accordant angles (less than 90 degrees). Such a drainage pattern tends to develop on gently sloping land with an impermeable clay subsoil. The drainage pattern is not influenced at all by variations in rock type or structure. The surface upon which it develops is entirely uniform.

Consequent streams

There are two major types of consequent stream. Some of them have courses that tend to follow the slopes and depressions that have been created by earth movements such as folding and faulting. A good example is shown in Figure 1.24(*b*)(*i*), where the land surface has been elevated to form a dome, and a set of four radiating streams, flowing down the slopes of the dome, has come into existence. A similar set of consequent streams could originate on the slopes of a newly formed volcanic cone. Figure 1.24(*c*) shows a rift valley that has been created by subsidence of the land between two parallel faults. Consequent streams flow to the centre of the rift valley from valleys that they have excavated in the fault scarps on each side. In Figure 1.24(*d*) earth movements have created a number of parallel anticlines and synclines. The synclines are occupied by consequent streams, flowing along their axes. Such streams are referred to as longitudinal consequents. Also, the folding has created slopes on each side of the longitudinal depressions. Streams flowing down these slopes,

Small-scale dendritic drainage patterns in a salt marsh creek

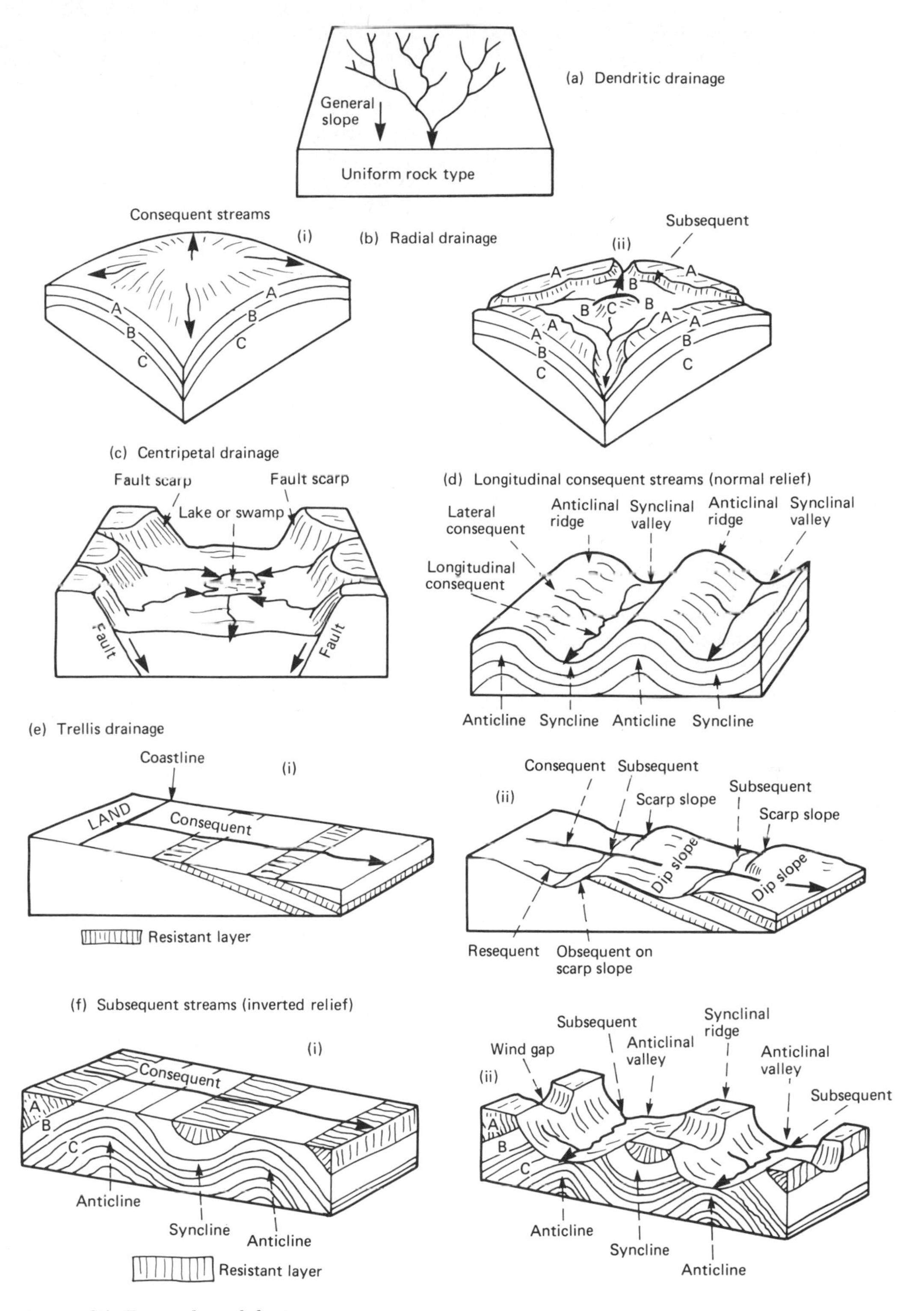

Fig. 1.24 Examples of drainage patterns

on the flanks of the anticlines, are called lateral consequents. They join the longitudinal consequents almost at right-angles. Figure 1.24(*d*) shows an example of 'normal relief'. The anticlines form the ridges of high ground and the synclines form the intervening valleys. Figures 1.24(*e*) and (*f*) illustrate a second type of consequent stream. In these examples the stream originates on the sloping land surface that is created when a former sea bed has been raised above sea level. It is important to notice that the courses of such consequents may be quite unrelated to the rock structures of the area. The consequent stream in Figure 1.24(*f*), for example, flows indiscriminately across anticlines and synclines.

Subsequent streams

The courses of subsequent streams are more often related to geological structure. They develop as tributaries of the original consequent streams and tend to excavate valleys in the outcrops of weaker rocks or along zones of weakness caused by faults or the joint pattern. In Figure 1.24(*b*) rock layers A and C are more resistant to erosion than layer B. When the crest of the structural dome has been removed by erosion all three rock layers are exposed at the surface (Fig. 1.24(*b*)(*ii*). Layer A forms an upland with steep scarp slopes facing inwards towards the centre of the dome, and layer C forms a small, dome-shaped upland in the centre. Layer B, on the other hand, has yielded more easily to stream erosion and forms a lowland at the foot of the scarp slopes. This circular lowland has been excavated by subsequent streams that are tributaries of the four original consequents, and which flow approximately parallel to the scarp slopes formed by layer A. Such lowlands are sometimes called 'strike vales' since they are parallel to the strike of the rock layers and are at right-angles to their dip.

Subsequent streams are also illustrated in Figures 1.24(*e*) and (*f*). In Figure 1.24(*e*) a consequent stream flows transverse to the outcrops of alternate layers of weak and resistant rocks on a former sea bed that has been exposed by uplift (Fig. 1.24(*e*)(*i*)). Figure 1.24(*e*)(*ii*) shows the possible development of the drainage pattern. Subsequent streams, at right-angles to the original consequent, have excavated parallel strike vales in the weak rock layers. These vales are asymmetrical in cross profile, having steep scarp slopes where the ground surface cuts across the resistant layers. The subsequent streams have two sets of tributaries. *Resequent streams* flow in the same direction as the dip of the strata and *obsequent streams* flow from the scarp slopes, in the opposite direction to the dip of the strata. These resequents and obsequents should not be confused with the lateral consequents shown in Figure 1.24(*d*). Resequents and obsequents cannot exist until there are subsequent streams. Figure 1.24(*f*) shows a more complex case. The original consequent stream flows across the outcrops of alternately weak and resistant rock layers that have been folded. Downward erosion by the consequent stream is restricted to some extent by the presence of the resistant layers in its course. However, subsequent streams that may develop along the outcrops of weak rock have no such restriction. They may therefore deepen their valleys relatively quickly and may become the dominant rivers of the area. The result might be that deep valleys are formed along the strike of the rock layers, in this case along the axes of the two anticlines. The consequent stream might cease to exist, but its former valley could be indicated by wind gaps, as shown in Figure 1.24(*f*)(*ii*). Such a relationship between topography and rock structure is referred to as 'inverted relief'. It is 'inverted' in the sense that the synclines (downfolds) form ridges and the anticlines (upfolds) form valleys. Figure 1.24(*f*) should be compared with Figure 1.24(*d*) which shows an area of similar geological structure whose drainage pattern is dominated by longitudinal consequent streams instead of subsequents.

Adjustment of drainage to structure

With the passage of time drainage systems tend to become adjusted to structure. Their plan increasingly reflects the underlying geological structure through the development of subsequent streams. An initial consequent drainage pattern may or may not reflect the geological structure. The pattern illustrated in Figure 1.24(*d*) quite clearly does, but the ones shown in Figure 1.24(*e*) and (*f*)(*i*) are quite discordant. This is because consequent streams tend to reflect the initial relief of the land surface and to have little relationship to the structure. Subsequent streams, however, by definition, follow lines or zones of structural weakness. As subsequent streams develop, therefore, the drainage pattern becomes more closely adjusted to structure. The consequent streams lose

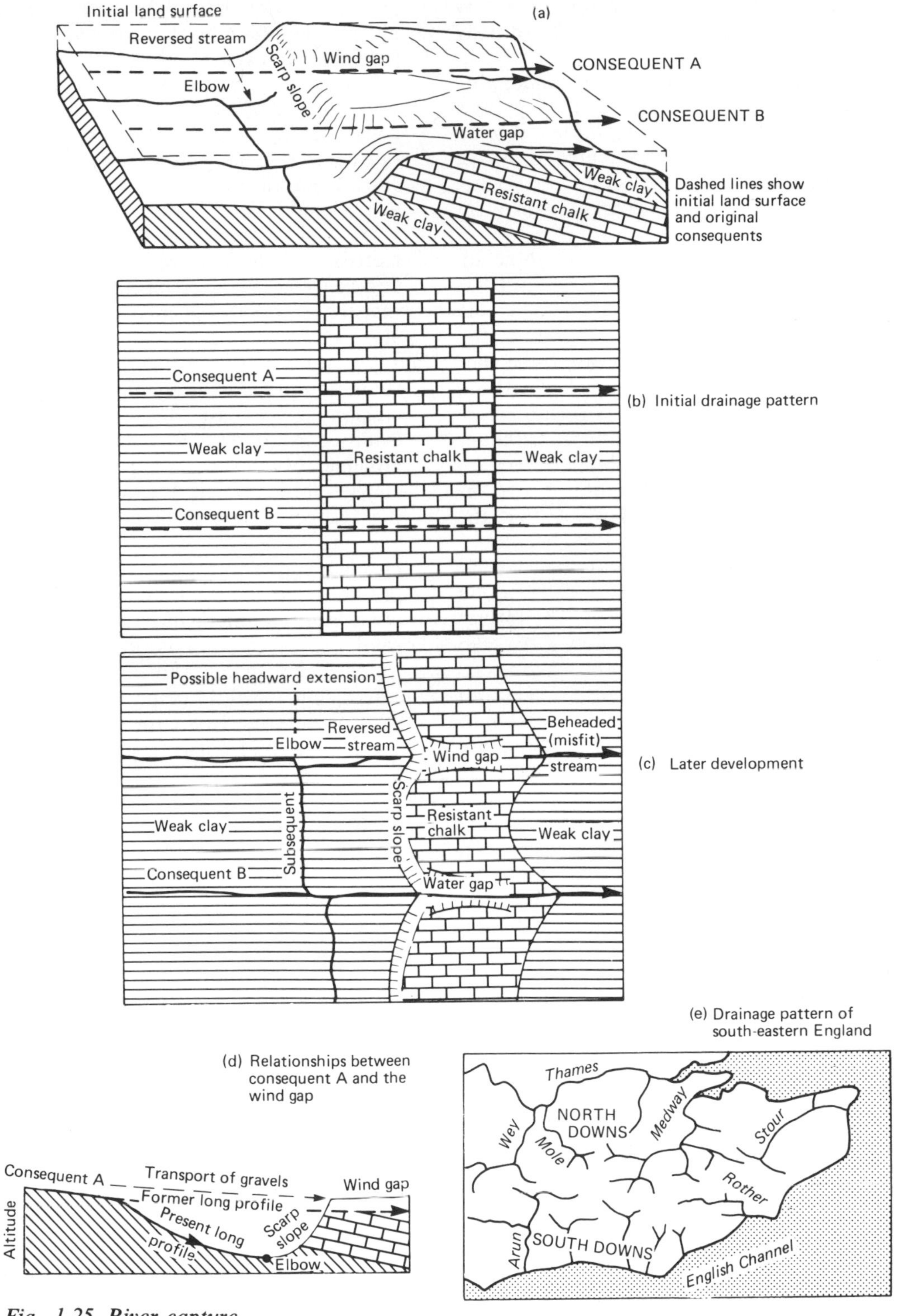

Fig. 1.25 River capture

their dominance often through being captured by newly developing subsequents. Eventually drainage patterns come to be dominated by subsequent streams and tend to reflect the underlying geology. The degree to which a drainage pattern has become adjusted to structure is therefore an indication of its age. Newly created drainage systems are unlikely to be adjusted to structure. Figure 1.24(*e*) illustrates the process of adjustment to structure.

Figure 1.25 shows this process in more detail. An area recently raised up above sea level has alternate layers of resistant chalk and weak clay forming parallel outcrops (Figs. 1.25(*a*) and (*b*)). Two consequent streams flow across these outcrops. Both these streams excavate valleys in both the clay and the chalk. Consequent B is the more vigorous and proceeds to develop a tributary along the outcrop of clay. This subsequent stream extends its source headwards until it reaches consequent A, which it 'captures', so that the upper part of A's course is diverted and becomes a tributary of consequent B. A right-angled bend (an elbow of capture) is created. The valley along the clay outcrop is further deepened. The former lower course of consequent A is now a dry gap (wind gap) at the crest of the chalk scarp and a small 'misfit' stream. This stream is clearly too small to have been responsible for the creation of the valley that it now occupies. It is sometimes referred to as a 'beheaded' stream, since its headwaters have been diverted to consequent B by the river capture that has taken place. It is possible that a completely new stream may flow from beneath the wind gap, down the scarp slope, to the elbow of capture. This is referred to as a 'reversed stream'. It is really an obsequent stream. Consequent A never did flow *up* the scarp slope to reach the wind gap. Consequent B, in contrast, has gained an extra supply of water through the capture of consequent A. Hence, it is able to deepen its water gap that crosses the chalk outcrop. As shown in Figure 1.25(*c*), the new subsequent stream may now extend headwards beyond the elbow of capture and may possibly capture the headwaters of other consequent streams.

All this seems very plausible in theory but problems arise when an attempt is made to interpret the history of a drainage network in the light of these ideas. It is not enough to search maps for elbows of capture, beheaded streams, and wind gaps. Attention has to be paid to comparative altitudes and features that can only be observed in the field. For example, if one is to conclude that a tributary of consequent B extended itself headwards by source recession and eventually captured consequent A, one would expect to find evidence that at some time in the past consequent B was at a considerably lower level than consequent A so that water could flow towards consequent B (Fig. 1.25). Also, if consequent B suddenly gained an increase in volume of water one would expect to find in its valley, river terraces or other evidence that the river had deepened its channel. Problems also exist in the identification of wind gaps (Fig. 1.25). Many hill ridges have depressions in their summits and these cannot all be attributed to river erosion in the past. Some kind of supporting evidence is needed. For example, it might be possible to study the long profile of the upper course of consequent A to judge whether or not the river could have passed through the wind gap. Such an extrapolation of consequent A's long profile is shown in Figure 1.25(*d*). In some cases a genuine wind gap has been identified by the study of sediments found on its floor. The presence in the wind gap of gravels composed of rock material which forms an outcrop in the upper course of consequent A would be good evidence that consequent A formerly occupied the wind gap.

Figure 1.25(*e*) illustrates the main features of the drainage pattern of the south-east corner of England. In this area the adjustment of drainage to structure has made considerable progress. The geological structure of this area resembles a dome, elongated from east to west. The North and South Downs are chalk uplands with scarp slopes facing inwards towards the centre of the Weald. Outcrops of the various strata run generally from east to west. From a study of the drainage networks it can be seen that the Medway has been most successful in developing subsequent streams and it appears that river capture may have occurred. Most of the rivers flowing towards the English Channel appear to have been less successful in developing subsequent tributaries.

Superimposed drainage and antecedent drainage

One of the reasons why there may be a discordance between the drainage pattern and the geological structure of the land surface over which it flows is that there has not been sufficient time for a major development of subsequent streams. The drainage system may have evolved on an overlying rock cover and, having removed this cover by erosion, the rivers now flow over a surface with a totally different structure. Alternatively the land surface over which the rivers flow may only recently have been elevated above sea level as in Figure 1.24(*e*)(*i*) and (*f*)(*i*). Both of these cases are referred to as superimposed drainage. Figure 1.26 shows an interesting example of the superimposition of drainage in the Hampshire Basin. Apart from an area in the north-west and another area in the east, the whole of this area has been raised above sea level comparatively recently. In the north-western area, which was not covered by the sea, the rivers Wylye, Nadder and Ebble flow generally parallel to the trend of the anticlinal axes. These rivers are well adjusted to structure. The Ebble, for example, flows along a synclinal valley like that shown in Figure 1.24(*d*). The Nadder, however, follows an anticlinal valley (Fig. 1.24(*f*)), flanked by inward facing chalk scarps. On the other hand the Avon cuts across the fold axes at right angles (Fig. 1.24(*f*)(*i*)), as do other streams further east, such as the Test and the Itchen. Even in the cases of the relatively young Test and Itchen, however, there appears to be a tendency for subsequent tributaries to develop near their sources, and a case of river capture may be suspected in the upper course of the Itchen (Fig. 1.26).

Another possible reason why a drainage pattern may be discordant to geological structure is that the rivers may be older than the rock structures over which they flow. In a few parts of the world, where rapid earth movements are known to occur, some rivers flow in deep gorges straight across high mountain ranges. Examples are the Brahmaputra (Tsangpo) and the Tista in the Himalayas, and the Columbia River that crosses

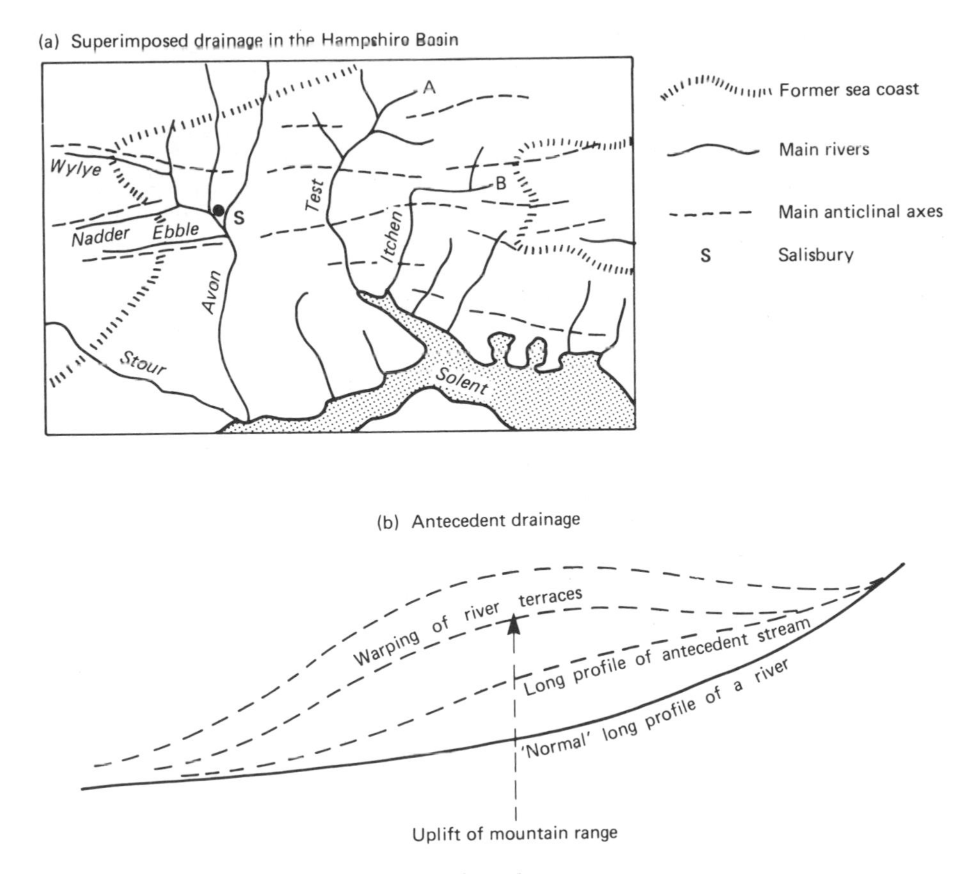

Fig. 1.26 Superimposed and antecedent drainage

the Cascades in north-west USA. In such areas rivers are able to erode vertically downwards fast enough to keep pace with the rate of uplift of the mountains. The rivers are therefore able to maintain their course across the rising mountain range. In some cases, however, the uplift of the mountain range results in the antecedent stream's long profile being convex upwards (Fig. 1.26(*b*)). Similarly, river terraces, indicating the former positions of the river, may be warped upwards by the rising mountain range.

RIVER CHANNEL FORM AND PROCESSES

Water in a drainage basin travels over the land surface and by underground routes to river channels which then convey it through the drainage basin and usually, eventually, to the sea. River channels also transport weathered rock debris derived from valleyside slopes (Fig. 1.13(*e*) and page 17) and other rock material eroded from river channels. The ability of a river to erode its channel and to transport rock debris depends upon its velocity.

The velocity of stream flow

The velocity of the flow of a stream is influenced by a number of different factors. One of these is the stream's *discharge*, which is defined as the amount of water that passes through a cross-section of the stream's channel in a certain amount of time. It is measured in cubic metres per second (cumecs). The hydrological equation $Q = AV$ means that the discharge (Q) of a stream is equal to the cross-sectional area of the stream channel (A) multiplied by the mean velocity of the stream (V). Hence, if a stream's discharge (Q) increases, there must also be an increase in either its cross sectional area or its velocity or both. In practice it is likely that both will increase and the stream will deepen and widen as well as increasing its velocity.

A stream's velocity is also influenced by the gradient of its channel. The steeper the channel the greater the stream's velocity. This is because stream flow is caused by the force of gravity and this is greatest in a vertical direction. The force of gravity along a sloping stream channel is proportional to the sine of the angle of slope. It is most effective in the case of a vertical stream channel ($\sin 90^\circ = 1.0$), as at a waterfall. With a 30-degree slope the gravity force is halved ($\sin 30^\circ = 0.5$). At slopes of less than 30 degrees the force of gravity is reduced very rapidly until it reaches zero when the stream channel is horizontal (Fig. 1.27).

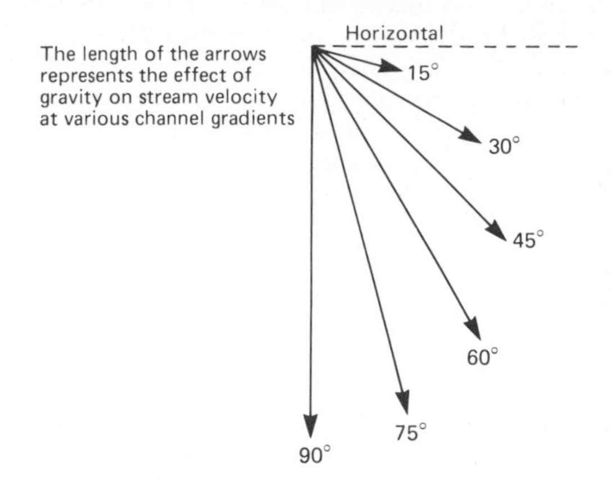

Fig. 1.27 The influence of gravity on stream velocity

The hydraulic radius of a stream channel is calculated by dividing its cross-sectional area by the length of its wetted perimeter. It is at the wetted perimeter that the velocity of stream flow is reduced by friction. Hence a large cross-sectional area and a small wetted perimeter (i.e. a high hydraulic radius) tends to be efficient and to favour a high stream velocity. Figure 1.28 illustrates this principle. In this figure, (*a*) illustrates the cross profile of a wide, shallow stream channel that has a large wetted perimeter in relation to its cross-sectional area, and hence a low hydraulic radius. Stream velocity is likely to be low. In Figure 1.28(*b*) the stream has the same cross-sectional area, but it has a smaller wetted perimeter. Its greater efficiency is indicated by the higher value of its hydraulic radius. In Figure 1.28(*c*) the stream has an even higher hydraulic radius; its semicircular wetted perimeter being even smaller than in (*b*). A semicircle is the ideal shape for the cross profile of a river channel in that it imposes least restriction on stream velocity. Figure 1.28(*d*) introduces a different aspect. It is exactly the same shape as the profile shown in (*c*), but its cross-sectional area is $16\ m^2$ instead of $8\ m^2$. However, although the cross-sectional area has doubled, the wetted perimeter has increased relatively little. Hence its hydraulic radius is considerably larger. The channel shown in (*d*) is therefore much more efficient than that shown in (*c*) merely because of its greater size. We can

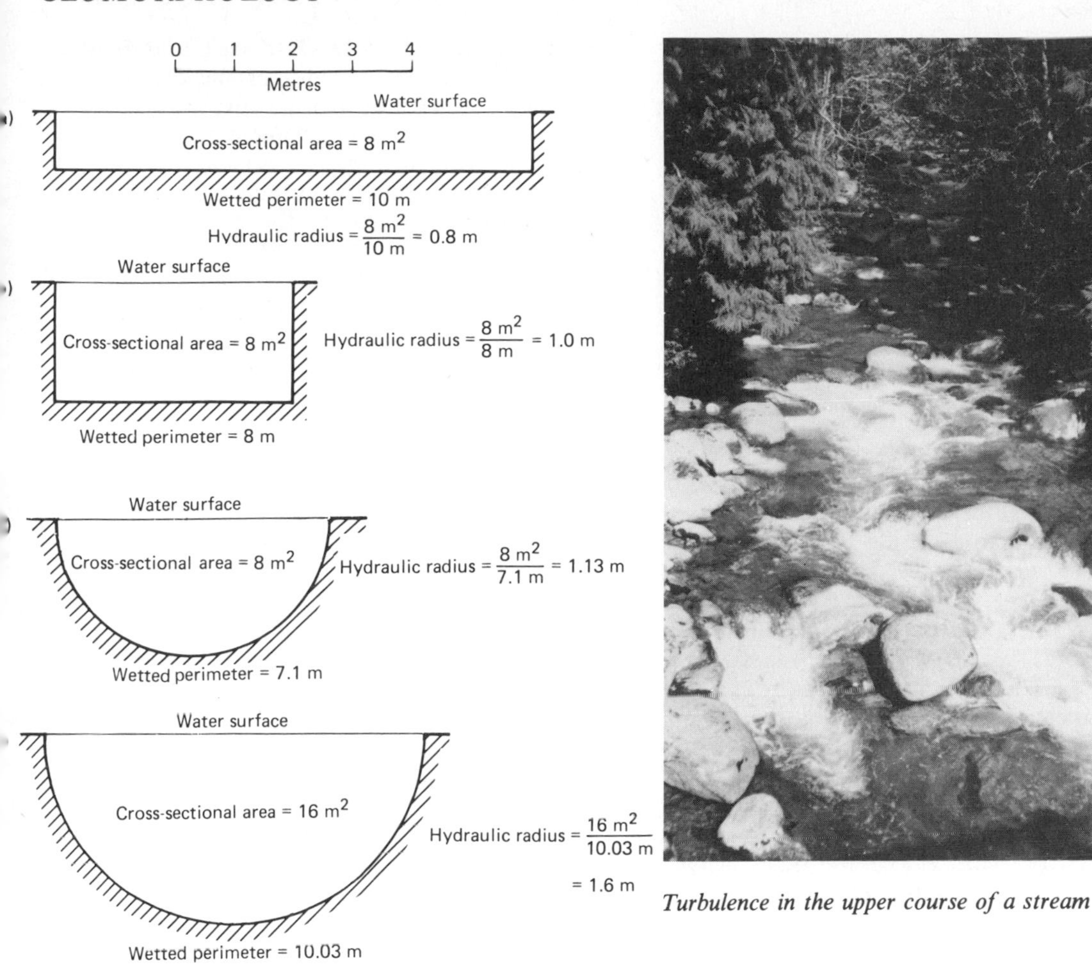

Fig. 1.28 The influence of channel size and cross profile on the hydraulic radius

Turbulence in the upper course of a stream

therefore conclude that stream velocity tends to increase as channel cross-sectional area increases.

Stream velocity is also strongly influenced by the roughness of the stream channel. The presence of boulders or irregular rock outcrops interferes with stream flow, causing turbulence and eddying which reduce the overall velocity of the stream. Streams whose channels are lined with sand or mud tend to flow faster.

Contrary to general belief, the average velocity of rivers tends to increase, or at least to remain constant, downstream from the source, despite the usual decrease in gradient. This is because, in a downstream direction, the size and the hydraulic radius of the stream channel tend to increase and also the wetted perimeter tends to become smoother. Stream velocity differs in different parts of a stream channel's cross-section. In a straight, symmetrical channel it tends to be highest in the centre, near the surface.

Transport by river channels

A stream transports its visible load of sediment in several different ways. Large rock fragments tend to roll or slide along the stream bed. If they are rounded in shape they may move comparatively easily, but if they are angular they may become wedged between other fragments, though in some cases they may slide over smaller fragments. Such movements are referred to as traction. *Traction* is usually most important near the source of a stream where river channel and valleyside gradients are steep, and valleyside slopes are capable of delivering coarse rock debris to stream channels. Smaller grains may be transported by *saltation*. In this case they are lifted bodily from the stream bed by turbulence and then they fall back again a short distance downstream. The finest grains of all may be transported in suspension in the body of the stream. The invisible load of dissolved materials is carried in solution. Generally, in a downstream direction, a river's load of sediment comes to consist of finer and finer fragments. This is because the particles transported by the river have been broken up by attrition and also because gentler valleyside slopes are only able to deliver material of finer calibre to the river channel.

The ability of streams to transport their load of sediment depends fundamentally upon their velocity, but a distinction should be made between capacity and competence. The capacity of a stream refers to its ability to transport a particular volume of sediment. It varies with the third power of the stream's velocity. Thus, if the stream's velocity doubles, its capacity increases by 2^3 (i.e. 8) times. Competence, however, is more important.

A river's *competence* is its ability to transport particles of sediment of various sizes and weights. It varies with the sixth power of the river's velocity. This means that if its velocity doubles, the river can transport particles 2^6 (i.e. 64) times heavier than before. This is because fast-flowing streams have greater turbulence and are therefore better able to lift particles from the stream bed.

Erosion in river channels

The competence of a stream greatly influences its ability to erode its channel. Erosion is most effective when the discharge is at a high level and the stream's velocity is high. In this case, the stream's competence will be great, as explained above. When a river lifts and transports particles from the stream bed it is performing erosion. But the river's hydraulic action can do more than this. If the floor and sides of the channel are composed of rocks which are poorly consolidated or well jointed, the hydraulic force of the flowing water may loosen blocks of various sizes and carry them downstream. This happens particularly when the river is turbulent.

In many cases, however, the channel's floor and sides are composed of rock that is too strong to be loosened by hydraulic action alone. In these cases channel erosion can take the form of abrasion. The river's load of sand, gravel, and boulders scours the floor and sides of the channel. The effectiveness of this process depends upon the mass of the moving rock fragments and also their velocity. Also, in highly turbulent streams cavitation increases the effectiveness of erosion. Shock waves are generated by the collapse of airless bubbles created by the turbulence of the water. If the stream is carrying a large load along its bed, vigorous vertical erosion can occur when discharge and velocity are at a high level. Cylindrical river potholes are created in the solid rock of the river bed, frequently at points of weakness created by joints. In these potholes loose stones are rotated, smoothing the walls and increasing the size of the pothole, so that eventually neighbouring potholes may join together to form a rock-walled gorge. Vigorous vertical erosion also takes place in the plunge pool at the base of a waterfall. Eventually, enlargement of the plunge pool can undermine the face of the waterfall and cause it to retreat upstream, thus creating a gorge by waterfall recession. Rivers can also perform lateral erosion by attacking the sides of the channel. If the channel sides consist of unconsolidated material this can be undermined so that it collapses into the channel and the debris is then transported downstream. When streams meander the more vigorous water flow on the outside of bends may undermine the bank and create a river cliff. Also the meanders themselves may migrate bodily downstream (page 40).

In addition to the mechanical erosional processes explained above, rivers are able to erode their channels by chemical solution. This occurs particularly in limestone areas (page 48).

Deposition in river channels

If the competence of a stream decreases its load of sediment begins to be deposited in the river channel. This happens when a decrease in the stream's velocity occurs, as a result of a reduction in its discharge, its gradient, or its hydraulic radius. A relatively small reduction in the stream's competence results in the deposition of only the coarser fractions of the stream's load, but a further reduction results in the finer fractions being deposited.

In upland areas with steep-sided valleys, tributary streams flow into the main valleys along very steep gradients and, at times, they can be heavily loaded with sediment derived from their valleyside slopes. When they reach the main valley floor there is a sudden decrease in their velocity and, often, in the hydraulic radius of their channels. Hence, their load of sediment (often of coarse calibre) is deposited to form an *alluvial fan* (or alluvial cone if it is very steep). This may be so permeable that the stream can often percolate through it. A series of such alluvial fans may form a continuous piedmont alluvial plain along the side of a major valley.

River deposition can take place within the river channel itself and also on the river's flood plain.

An example of deposition within the channel is the creation of a *braided channel.* Braiding involves the deposition of banks of sand or gravel, within the main channel, between which the stream divides into several separate channels which repeatedly rejoin and redivide. The load of rock debris is too large to be transported by a single channel. At times of high discharge the banks may be undermined so that they collapse into the channel. Thus the channel tends to widen, but deposition within the channel tends to make it shallower. Thus, the wetted perimeter increases with perhaps little change in the cross-sectional area. Hence, the hydraulic radius is low. To compensate for this a braided river channel is usually steeper than, for example, a meandering channel. Braiding is common in semi-arid areas where there is a great deal of rock waste and only a low rainfall and stream discharge. It is also found in the channels of meltwater streams flowing from glaciers. Here, much rock waste is supplied as the glacier melts, and the stream's discharge varies in relation to the rate of melting of the glacier.

Both erosion and deposition take place in a meandering stream. In a meander (Fig. 1.29(*a*)) the flow of the stream is fastest along the concave banks and the fastest flow of all tends to occur near the downstream end of a concave bank. Vigorous undercutting tends to occur along these concave banks and eroded material slumps into the river. Some of this sediment is transported to the opposite convex bank by a subsurface flow of water and accumulates to form a point bar (Fig. 1.29). From the downstream end of the concave bank eroded sediment tends to be transported directly downstream and deposited on the next convex bank on the same side of the river. The result of these processes is that, in meanders, the concave banks tend to retreat and the lower, flatter convex banks gradually advance. The original valleyside slopes are replaced by almost level point bars, deposited by the river, and they become the river's flood plain. Figure 1.29(*b*)(*i*) illustrates this process. If the curves of the meander continue to be accentuated in this way the river will eventually cut through the meander neck and take a direct course across it, bypassing the former meander loop, which becomes a cut-off, possibly occupied by a temporary lake which gradually silts up. Meanders also tend to migrate bodily in the direction in which the river flows. This process is illustrated in Figure 1.29(*b*)(*ii*). The result is that a flood plain bounded by bluffs (formerly river cliffs) is created. Usually the two processes, meander accentuation and downstream migration, occur simultaneously.

The flood plain is the generally level ground bordering the river channel that is covered when the river floods. It usually contains a few depressions such as former cut-off channels, and it is composed largely of point bars built up by former meanders. When it floods, the river deposits sediment over its flood plain. Often this is finer silt or clay which covers the coarser alluvium of the point bars. The level of the flood plain tends to rise towards the river channel, forming natural levees. When the river floods it may cover the whole of the flood plain but the water still flows fastest along the line of the river channel because here the hydraulic radius is greater than in the case of the shallow flood water that covers the flood plain. Hence, along the margins of the river channel there is a sudden decrease in both competence and capacity. This causes deposition to occur, beginning with sediment of coarser calibre. Sediment of finer calibre can be deposited more widely over the flood plain.

Deltas are formed when rivers enter standing water in the sea or a lake. Their velocity decreases and the consequent reduction in competence leads to the deposition of sediment. Deltas are the continuations of the flood plains of rivers. The formation of a delta is dependent upon the relationship between the rate at which the river delivers sediment to the coastline and the rate at which this sediment is carried away by the action of waves and currents. If a river delivers very large amounts of sediment to its mouth it is possible for a delta to be built even though there are strong sea currents and a large tidal range. In North America the Colorado River has built a delta into the Gulf of California despite a tidal range greater than that of many parts of the coast of England. The shape of a delta depends upon the relationship between sedimentation by the river and marine processes of deposition and erosion. If wave erosion and current action are strong the delta may be *cuspate* in shape or merely the infilling of an inlet (Fig. 1.30). A *lobate* delta (Fig. 1.30(*c*)) may indicate a strong influence of river deposition. The river channel has divided into several distributaries and sedimentation has taken place along each of these. Lagoons may have been formed as distributaries have converged. The Mississippi has built an outstanding example of a lobate delta that advances seawards at something like 60 m per year. Smoothly curving arcuate deltas have been influenced to a greater extent by waves and currents in the sea. Waves and currents have either removed the delta lobes or prevented their forma-

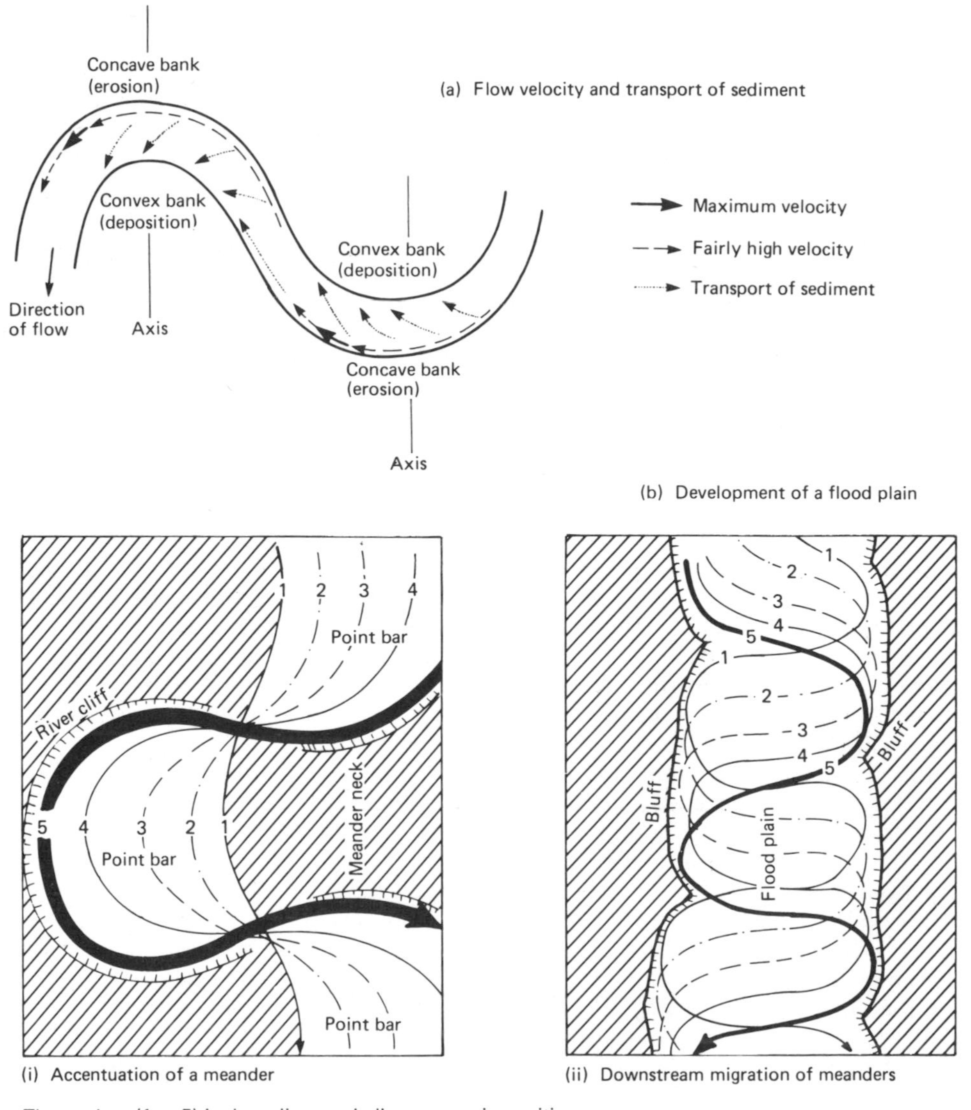

Fig. 1.29 Meanders

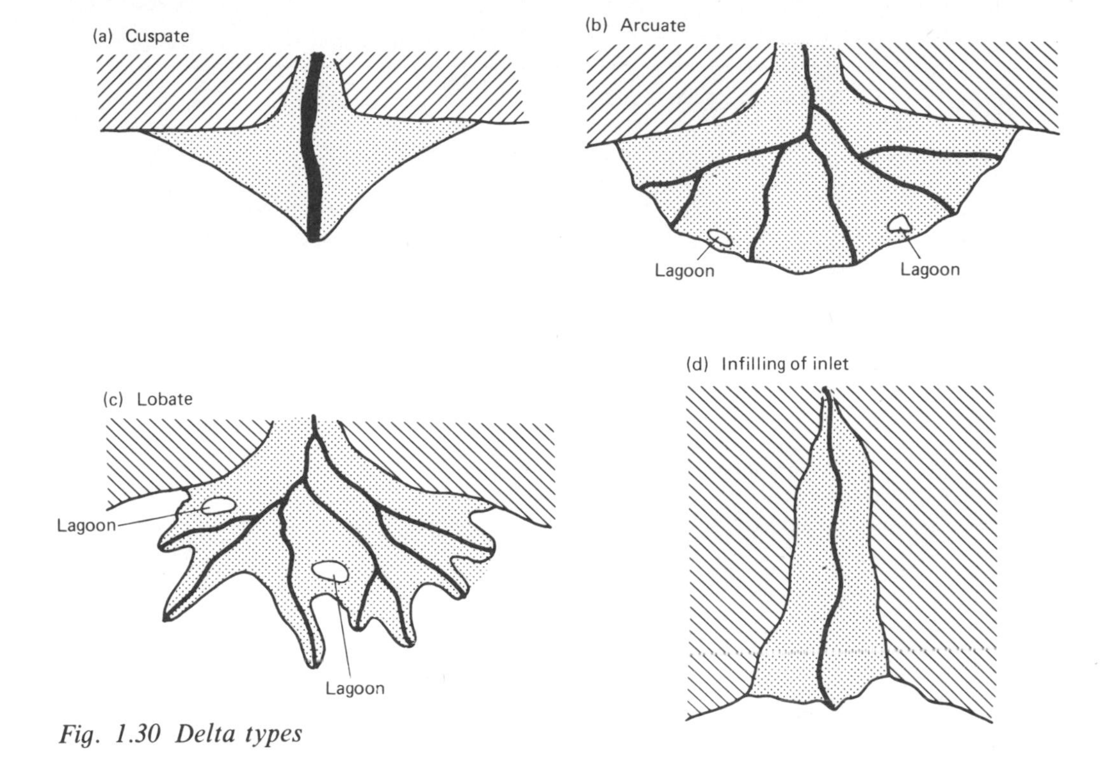

Fig. 1.30 Delta types

tion. Spits and bars may have developed along the seaward margin of the delta, again forming lagoons (Fig. 1.30(*b*)). The Nile and the Rhine have deltas of this type. The world's deltas appear on maps to be very much smaller than they really are because the melting of the ice sheets at the close of the Ice Age caused a world-wide rise of sea level. The Nile delta, for example, rises at least 1000 m from the Mediterranean sea bed and extends at least 100 km to the north of its present coastline.

River traces

Very few rivers flow in a perfectly straight line for any considerable distance. If they do, they are probably flowing down steep slopes or they are strongly influenced by the presence of zones of weakness, such as joints or faults, in the underlying rocks. Meandering river channels, however, are very common.

If a straight river channel has a bedload of mixed calibre, such as a mixture of sand and gravel, it tends to have an undulating bed, consisting of alternating pools (deeper areas floored by sand) and riffles (accumulations of gravel forming shallower areas). Water flows rapidly over the riffles and more slowly in the pools (Fig. 1.31(*a*)). Such a sequence of shallows and deeps in a stream does not seem to be stable, since alternate pools tend to migrate to opposite sides of the channel (Fig. 1.31(*b*)), thus causing the *thalweg* (the line of greatest depth) to swing from one side of the channel to the other. This appears to be the beginning of a process that creates meanders. Meandering river channels also have deeps and shallows. The deepest parts are located near the concave banks. The shallows (riffles) are situated near the *points of inflection* (cross-over points), where the direction of curvature changes, and the river channel is comparatively straight (Fig. 1.31(*c*)). It has therefore been suggested that meander-forming processes operate in straight river channels and that this is why so few river channels are straight. One explanation that has been suggested for this is that streams tend to develop a 'steady state' in which the stream's energy is used at a uniform rate along its channel. In a straight channel much more energy is used as the water flows over a riffle than when it passes through a pool (Fig. 1.31(*d*)). In a meander a greater proportion of the stream's energy is used in

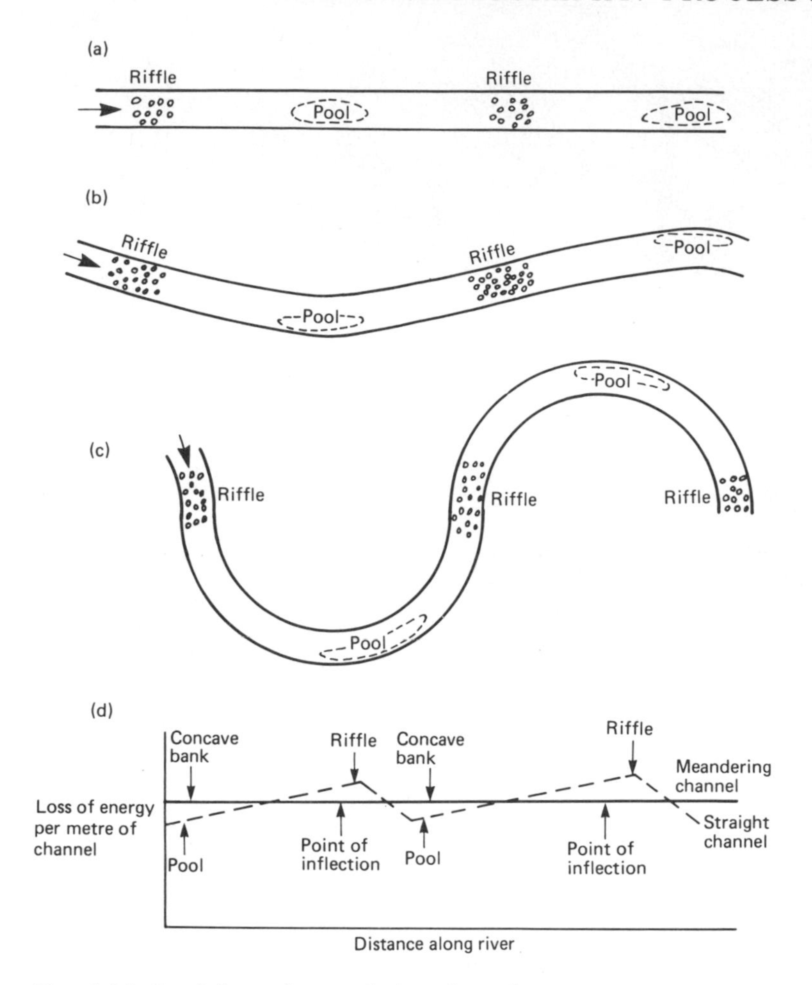

Fig. 1.31 Straight and meandering channels

the pool because here the channel is curved and erosion of the concave bank is likely to be taking place (Figs. 1.31(*c*) and (*d*)). At the points of inflection (riffles) there is a high loss of energy as a result of the shallow water (Fig. 1.31(*c*)), but here the channel is comparatively straight so little energy is lost as a result of curves in the channel. Thus, the rate of energy loss in a meandering channel is likely to have fewer fluctuations than that in a straight channel (Fig. 1.31(*d*)).

Certain quantitative regularities tend to exist in meanders. Most of the dimensions of a meander are related to the width of the river channel. Some of these relationships are illustrated in Figure 1.32. In this figure, for example, the radius of curvature of the meander is 2.5 times the channel width which is rather less than the average of 3.0. The valley length of a meander is usually between 7 and 11 times the channel width, but the channel length measured along the curves of the river is usually between 11 and 16 times the channel width. The corresponding values in Figure 1.32 are 10 and 15.7. Points of inflection tend to be spaced at between 5.5 and 8 times the channel width. In Figure 1.32 the figure is rather high at 7.85. The *sinuosity* of a meander is calculated by dividing the channel distance between two points of inflection by the straight-line distance. It shows the extent to which the meander deviates from a straight line. It tends to average about 1.5, slightly less than in Figure 1.32. The *amplitude* is indicated in Figure 1.32. This characteristic correlates extremely

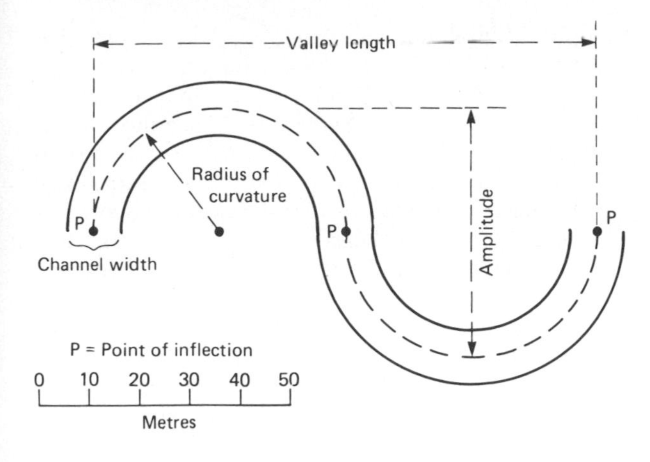

Summary of characteristics
Channel width = 10 m
Radius of curvature = 25 m
Valley length = 100 m
Channel length = 157 m
Spacing of points of inflection = 78.5 m
Amplitude = 50 m

Relationships

$$\frac{\text{Radius of curvature}}{\text{Channel width}} = 2.5$$

$$\frac{\text{Valley length}}{\text{Channel width}} = 10$$

$$\frac{\text{Channel length}}{\text{Channel width}} = 15.7$$

$$\frac{\text{Spacing of points of inflection}}{\text{Channel width}} = 7.85$$

$$\text{Sinuosity} = \frac{78.5}{50} = 1.57$$

Fig. 1.32 Quantitative aspects of a meander

poorly with channel width. The strongest influence on amplitude appears to be the varying resistance of the river banks to erosion.

The long profile of a river

Rivers have been described as examples of 'open systems' tending towards a 'steady state' or 'dynamic equilibrium'. An 'open system' is one that receives inputs of matter and energy from its environment and produces outputs that return to the environment. 'Dynamic equilibrium' means that the system achieves a balance between its inputs and its outputs. There is no tendency for progressive long-term change to occur. If equilibrium is temporarily disturbed, forces are set in motion to restore it.

The shape of a river's long profile basically depends upon the relationship between the river's discharge and velocity and the sediment load that is being transported. A tendency seems to exist for each part of a river's course to adjust itself so that it is just able to transport a certain amount of sediment of a certain calibre. When the total energy of the river at bankfull discharge (with its channel completely filled with water) is just sufficient to transport its water content and its load of sediment the river is said to be in equilibrium (graded).

Ideally the long profile of a river that has attained equilibrium has a gradual decrease in slope from its source to its mouth (Fig. 1.33(*b*)), and thus forms a concave-upward curve. This curve tends to resemble that which would be produced by a straight line drawn on semi-logarithmic graph paper (Fig. 1.33(*a*)). There are several reasons for this. In the headwaters, the relatively shallow stream has a low hydraulic radius, and large rock fragments eroded from the channel or delivered to the channel from the valleyside slopes cause the stream bed to be very rough. In order to transport such a coarse calibre load under these conditions the river's gradient must be steep. Contrary to general opinion, the river's velocity is not greatest in its headwaters. In fact river velocity tends to remain constant, or to

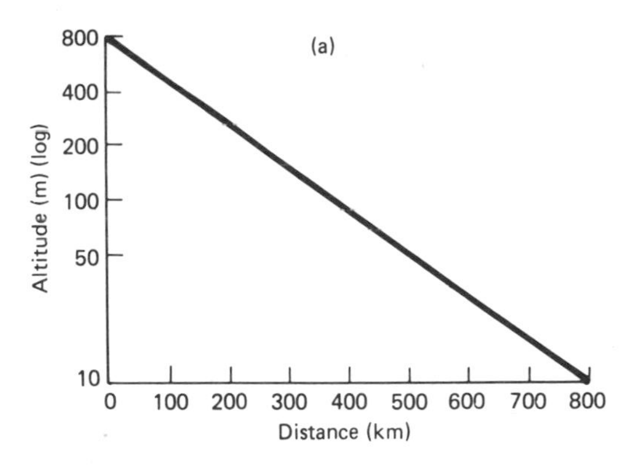

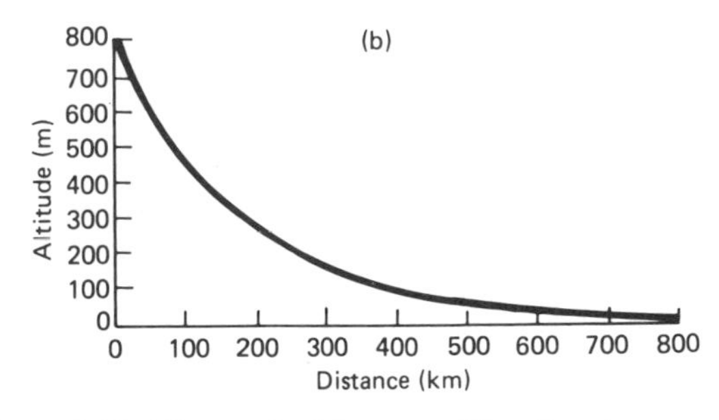

Fig. 1.33 Theoretical long profile of a river

increase slightly, in a downstream direction. Also, in a downstream direction the river's discharge increases as it receives tributary streams, and its hydraulic radius increases as it becomes deeper. In addition, its bed roughness decreases and its sediment load becomes of finer calibre. These factors would tend to increase the river's velocity in a downstream direction, but they are counteracted by the decrease in channel slope.

The normal concave-upward long profile of a river seems, therefore, to represent a relationship between the river's ability to perform the work of transport and the calibre of the material that is transported. In fact, examples of rivers with long profiles exactly like that shown in Figure 1.33(*b*)) rarely, if ever, occur. Irregularities in river long profiles can occur for a variety of different reasons.

The long profile of a river flowing over a newly-uplifted land surface seems certain to have been irregular, with steep reaches (rapids and waterfalls), level reaches and hollows that would be occupied by lakes. Such a river cannot be in equilibrium. In the steep reaches the river would tend to have a surplus of energy over and above that needed to transport its water and sediment content. It would therefore erode its channel and thus decrease the channel's slope. In the reaches with gentler gradients the river's energy would tend to be insufficient to transport its load of sediment, so deposition would occur, so as to steepen the channel slope. This principle is illustrated in Figure 1.34(*a*). Here, the resistant rock layer forms a temporary base level below which the upstream reach of the stream is unable to lower its channel. Erosion of the resistant layer tends to reduce the river's slope. On the weak rocks the

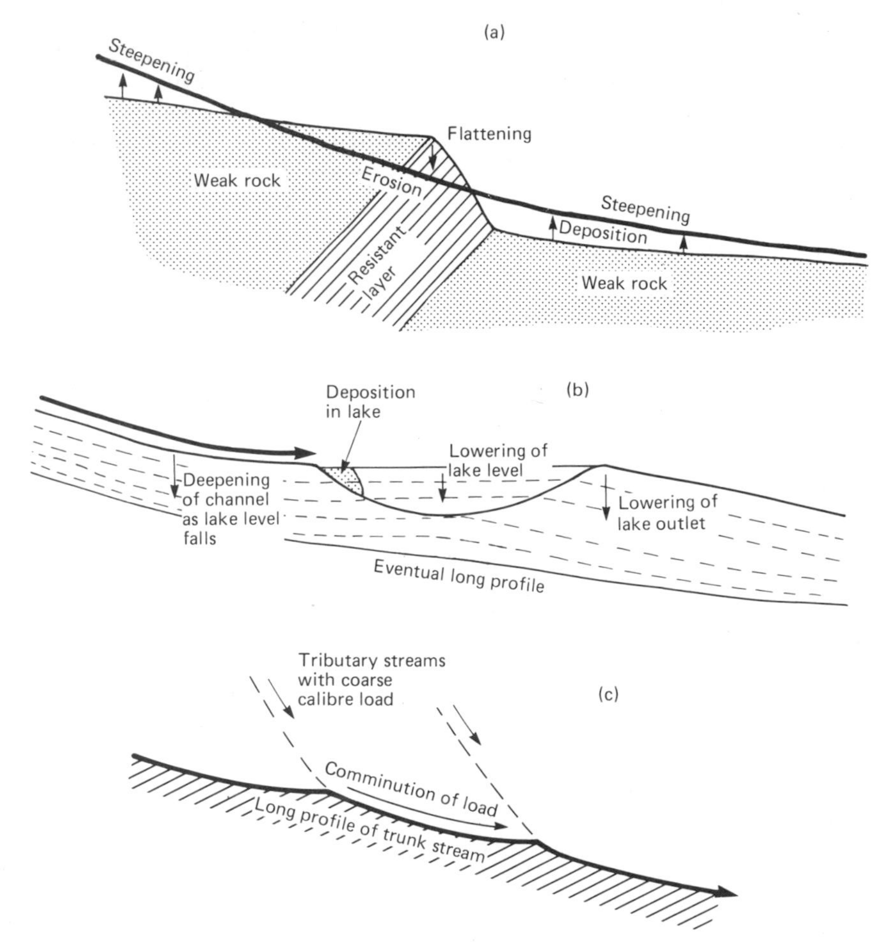

Fig. 1.34 Influences on the long profile of a river

river's energy may be insufficient to transport its load, so deposition occurs, thus steepening the gradient.

A lake has a rather similar influence on a river's long profile. The lake forms a local base level for the part of the river that lies upstream. Sediment transported along this part of the river is deposited in the lake since the lake water has a negligible velocity. Gradually, the lake will fill with sediment, but, of course, the level surface of a filled-in lake could not be a profile of equilibrium. We also have to take into account the stream that flows out of the lake. This stream receives very little sediment from upstream. Most of this has been trapped in the lake. Hence, it seems likely that the outlet stream will have more energy than it needs to transport its load. It will therefore tend to degrade its channel (Fig. 1.34(*b*)).

Tributary streams can also have an influence on the long profile of a river. They may be transporting a sediment load of a different calibre from that of the major stream. A tributary from a nearby mountainous area, for example, may be carrying a load of very coarse calibre down a steep gradient. This is likely to cause the trunk stream to steepen its profile below the confluence. Further downstream the coarse calibre load may be comminuted (reduced to smaller fragments), so that the trunk stream's gradient gradually flattens (Fig. 1.34(*c*)).

In some cases the type of rock that forms the valley sides and that underlies the river channel may influence the long profile. Shales or clays tend to produce a sediment load of very fine calibre, and limestones produce very little visible load. Hence, one might expect a river to have a low gradient when passing over such rocks. Sandstones, on the other hand, produce coarser, more resistant sediments, so one would expect the long profile to be steeper. One would not expect to see very sudden changes in gradient, however, since a bed load of sandstone fragments can be carried a considerable distance downstream before the fragments are comminuted by attrition.

Irregularities in rivers' long profiles may also occur as a result of a change in their base level, which is the level below which it is impossible for them to lower their channels by erosion. The base level for streams that flow into the sea is sea level. For streams flowing into enclosed lakes that have no outlet the lake surface is the base level. For example, the base level of the River Jordan is the surface of the Dead Sea, some 400 m below sea level. The base level for a tributary stream is the level of its confluence with the trunk stream. No river can degrade its channel any significant distance below its base level. Most rivers flow eventually into the sea, and sea level (their base level) may change from time to time. These changes may be positive (a relative rise of sea level) or negative (a relative fall of sea level). Such changes of sea level may be of either eustatic or isostatic origin. An eustatic change of base level occurs when the level of the sea (rather than the land) changes. It may do so as a result of the advances and retreats of ice sheets that have taken place from the Ice Age onwards. When ice sheets advance, precipitation falling as snow fails to return rapidly to the oceans and remains trapped in the ice sheets for many years. Thus, much water that is evaporated from oceans cannot return to them. Hence, a fall in sea level occurs. When the ice sheets do retreat, in an interglacial stage, sea level tends to rise again. An isostatic change of base level occurs as a result of a change in the level of the land rather than the level of the sea. Isostatic changes tend to be very slow indeed. The area surrounding the Baltic Sea is still rising slowly as a result of the removal of the weight of the ice sheet at the end of the Ice Age.

A positive change of base level tends to flood the lower parts of river courses, turning them into estuaries. Rivers then gradually fill these estuaries with sediment, thus producing alluvial flats resembling river flood plains, but their origin is quite different.

A negative change of base level can have a much greater effect upon relief. In this case the extension of the river's course over the former sea bed may have a steeper slope than the lower course of the river (Fig. 1.35(*a*)). If the river then deepens this part of its course a knickpoint may form which gradually retreats upstream, successively steepening each section of the river in turn and eventually lowering the whole profile. This influence also extends along the tributaries (Fig 1.35(*b*)). In a river system where this has occurred it may be possible to project the former long profiles of the major stream and its tributaries and discover that they are related to raised beaches on the coastline that indicate a former higher sea level. This process tends to create certain distinctive landforms. Knickpoints, for example, may exist in the form of waterfalls or rapids, but these features may also exist simply because of outcrops of resistant rock in river channels. A deepening of the river channel along the course of existing meanders can produce incised meanders (Fig. 1.35(*c*)). Two types of incised meanders are usually recognised. If the cross-profile of the winding valley is symmetrical they are known as intrenched meanders. Frequently, however, as in the case of flood plain meanders, an element of lateral erosion exists and this causes the valley side to be steeper on the concave bank of the river then on the convex side. These are known as ingrown meanders. It is even possible for the river to bypass a meander loop as it does in the formation of a flood plain cut-off, leaving an abandoned incised meander.

Another distinctive landform can be created as the river deepens its channel: it may leave its former flood plain as a fairly level terrace on each side of the valley and create a new flood plain at a lower level. The older flood plain is known as a river terrace (Fig. 1.35(*d*)). Further periods of downcutting by the river may create a 'staircase' of river terraces on each side of the valley. Such terraces tend to occur at similar heights on both sides of the valley. They are described as 'paired'. River terraces can sometimes be projected to the sea coast and correlated with raised beaches that indicate former sea levels.

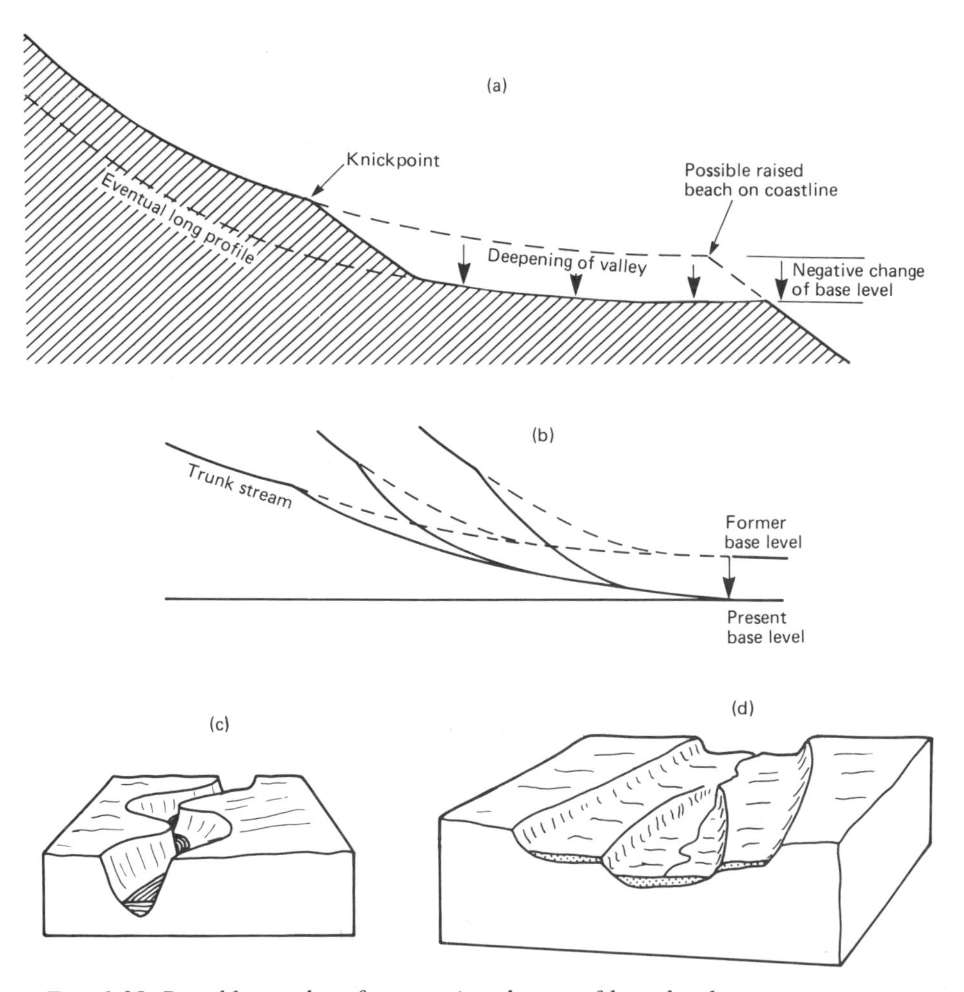

Fig. 1.35 Possible results of a negative change of base level

A stream flowing into a swallow hole. The main sink of Gaping Ghyll, Yorkshire

Limestone slopes at Malham Cove

1.4 The influence of rock type and structure

The relief of the land surface is the product of the interaction between external processes, such as weathering, mass wasting and river action, and the characteristics of the underlying rocks, including the folding and faulting that have affected them. Volcanic activity also produces a variety of landforms which are subsequently modified by weathering and erosion.

THE INFLUENCE OF ROCK TYPE

Rocks differ greatly from one another in respect of their hardness, their solubility, their porosity and permeability, and their structures such as jointing, cleavage and stratification. Weathering processes are able to attack the rock mass along joints and cleavage and bedding planes. Porous and permeable rocks tend to resist erosion because they have few surface streams. Hence, a weak rock such as chalk, although soft, is porous, and often forms relatively high ground and steep escarpments because it occurs adjacent to weak, impermeable clays that are comparatively easily excavated by river action. Igneous and metamorphic rocks, such as granite, gabbro and quartzite are hard and coherent and therefore resist erosion by rivers even though they are relatively impermeable. Smaller igneous intrusions, such as sills and dykes, often form escarpments and ridges as the weaker sedimentary rocks, into which they were intruded, are attacked by erosion. It is not so much the absolute hardness of a rock that influences the relief as its hardness in comparison with neighbouring rocks. Thus, for example, Carboniferous limestone usually forms relatively high ground but, if it occurs next to a more resistant rock, it can form relatively low ground.

Limestone landforms

Three common types of limestone in the United Kingdom are the Carboniferous limestone, in the Craven district of Yorkshire, the Peak district and the Mendips, chalk in the Chilterns and the North and South Downs, and the Jurassic limestone in the Cotswolds. These limestones consist mostly of calcium carbonate which is susceptible to solution by water containing carbon dioxide. They have very little surface water, so there is little stream erosion. Rain water mostly percolates through pore spaces (in chalk) and along joints and fissures (in Carboniferous limestone). These limestones therefore tend to form relatively high ground. In the Craven district the Carboniferous limestone usually forms a plateau at the foot of slopes composed of the less permeable Yoredale beds which are capped by a layer of impermeable Millstone Grit (Fig. 1.36(*a*)). The limestone here is hard and well cemented and produces landforms of the type known as *karst*. Rain water can percolate down through joints and this produces clints and grikes (page 10) by solution weathering. Streams flowing down from the Yoredale beds frequently sink into swallow holes, created by solution, soon after they reach the limestone. A dry valley beyond the swallow hole shows the former course of the stream (Fig. 1.36(*b*)). Upstream of such a swallow hole the stream may have continued to deepen its course, thus producing a blind valley (Fig. 1.36(*b*)). Some dry valleys may be related to the deepening of the main valleys by the major streams. This causes the water table to fall, so that smaller streams cannot maintain themselves (Fig. 1.36(*c*)). The underground streams reappear at the surface as springs or resurgences near the junction of the limestone with the underlying impermeable rocks (Fig. 1.36(*d*)) at the foot of the steep limestone scars. These scars frequently take the form of free faces with constant debris slopes (Fig. 1.13(*b*)). The plateau surface also contains a variety of closed depressions formed either by solution from the surface or by the collapse of an underground water course. Once a river has gone underground its course bears no relationship whatsoever to the relief of the ground surface. It is quite impossible to link a particular swallow hole with a particular resurgence on map evidence alone. In some cases an underground stream actually crosses the course of a surface stream or another underground stream. At a certain depth is the water table. Its form depends upon the permeability of the limestone. If the joints are small it tends to dome upwards under hills, but it is flatter if water can percolate more freely through the joints. Limestone caverns are believed to have been caused through the limestone being dissolved by water percolating laterally somewhere near the water table. As major rivers have deepened their valleys the water table has been lowered, thus leaving dry caverns at different levels (Fig. 1.36(*e*)). Research has established a correlation between cavern levels and pauses in the process of valley deepening by major streams. Once the water table has sunk it becomes possible for travertine (calcium carbonate) to begin to accumulate in caverns in the form of stalactites and stalagmites and other more complex formations.

Chalk is a softer type of limestone, with few well-marked joints. It has few surface streams because of its porosity, and it tends to form uplands rising above adjacent valleys excavated in soft impermeable clays. There are few surface depressions caused by solution, though some do exist in river channels. It is not compact enough to allow caverns to develop. The typical landform produced by a layer of chalk is the cuesta (Fig.

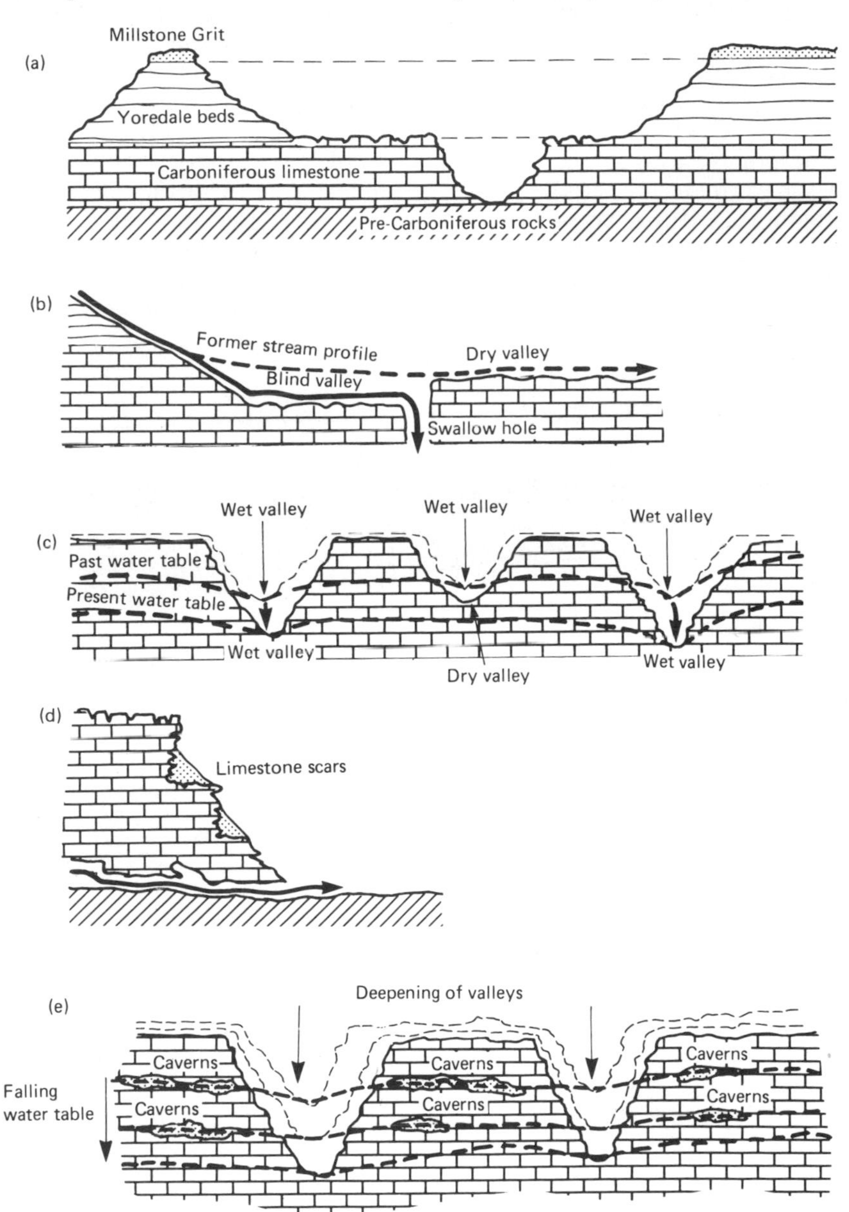

Fig. 1.36 Carboniferous limestone landforms

1.13(*c*)) with a steep scarp slope cutting across the strata and a gentle dip slope parallel to the strata. In some cases the gentle slope may not be parallel to the dip of the strata, so it can be referred to as a back slope. A chalk landscape as a whole is characterized by smooth convex curves without free faces. Since the chalk is porous, rain water percolates downwards beneath the water table which reaches a higher level under the hills than under the valleys (Fig. 1.37(*a*)). The water table often reaches the surface near the base of the scarp slope, where the underlying clay reaches the ground surface, and springs are formed. The water table also often reaches the surface in the valleys of the dip slope. These springs are capable of performing erosion. They have frequently eroded headwards (spring sapping) to create re-entrants in the scarp slope. Also, they have often produced a marked break of slope in the long profile of dip-slope valleys. In some areas huge rounded hollows (coombes) on the scarp slope appear to have little relation to spring sapping. These could have been created by freeze-thaw processes in periglacial conditions during the Ice Age. Frequently a mass of chalk sludge can be seen to have accumulated at the scarp foot near such a coombe. Although at the present time the dry valleys are above the level of the water table they appear to have been excavated by river erosion. One suggested explanation is that in interglacial or post-glacial times the climate was wetter and this caused the water table to be higher than at present, so that streams could flow in the valleys that are now dry. Another suggestion is that streams could flow over the frozen ground under periglacial conditions during the Ice Age. A further suggestion is that the dry valleys are the result of the retreat of the scarp slope. This is illustrated in Figure 1.37(*b*). The recession of the scarp slope causes the level of the water table to fall. Unless the streams can deepen

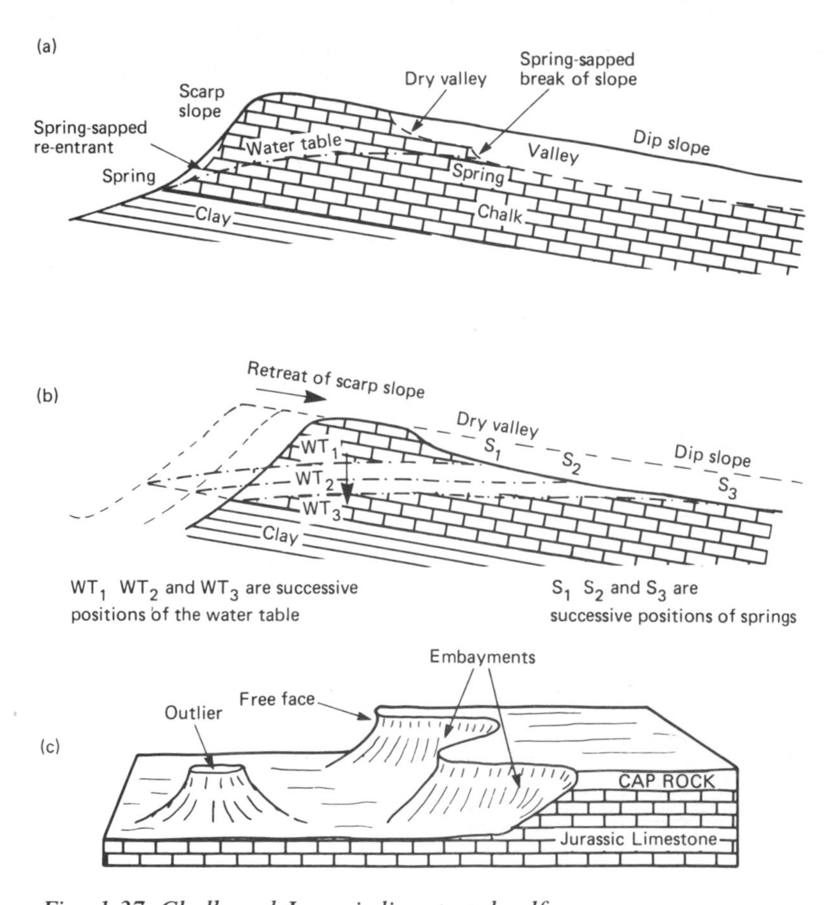

Fig. 1.37 Chalk and Jurassic limestone landforms

Devil's Dyke, a dry valley in the scarp slope of the South Downs

The scarp slope of the Cotswolds near Birdlip

their valleys at the same rate as the water table falls, their sources (springs) will migrate downslope and leave the upper parts of their valleys dry.

Jurassic limestone is harder than chalk but still not hard enough to produce a karst landscape. It does have a few surface streams in relatively widely spaced valleys. East and north-east of Gloucester, in the Cotswolds, the limestone is almost horizontally bedded and prominent, steep scarp slopes are sometimes capped by a free face of bare rock. The scarp slope here has been considerably dissected by rivers eroding into it headwards. This has created a number of embayments. Also, in places, as the scarp slope has been worn back irregularly, small steep-sided outliers have been left well to the west of the main escarpment (Fig. 1.37(*c*)).

Landforms of igneous rocks

In Devon and Cornwall a series of upland areas extending from Dartmoor to Land's End is composed of large granite intrusions in the form of bosses. These must have been dome-shaped originally, when they were intruded, and, to a great extent, they have kept their original shape. They have successfully resisted erosion by rivers and their detailed landforms are probably mainly the result of a period of chemical weathering in the past, when the granite tors were shaped (page 10, Fig. 1.8(*b*)). Probably, also at this time, more general weathering took place which was most effective in the areas where the joints were most closely spaced. Since then, it appears that river action has had little effect in areas of undecomposed granite, but decomposed granite has been removed to create a series of marshy basins and the cross profiles of river valleys are frequently wide and gently sloping. In places, the rivers cross outcrops of relatively unweathered granite. Here they have tended to erode vertically and produce narrow gorges. Surrounding the granite boss is a metamorphic aureole, an area where the pre-existing rocks were altered during the intrusion of the granite. Streams flowing from the granite have frequently excavated deep valleys here. It appears that the granite tors have also been influenced by processes that were in operation during the Ice Age. Large rocks appear to have fallen from the tors and to have moved downslope under periglacial conditions. Frequently a 'stream' of such rocks can be seen running from a hillside tor to a nearby valley floor.

Large masses of igneous rocks now form very contrasting landscapes. In the Isle of Skye there is a striking contrast between the rounded granite mountains and the jagged profiles of the Black Cuillins which are composed of gabbro.

Sills, extensive horizontal sheets of igneous rocks which have been intruded between the layers of sedimentary rocks, often cover large areas. They are sometimes more resistant to erosion than the sedimentary rocks into which they have been intruded, so they tend to form valleyside benches and waterfalls in river courses. The Whin Sill covers a large area of northern England and is responsible for striking waterfalls. Volcanic necks, composed of the lava which solidified in the vent of an ancient volcano, form striking, steep-sided, often isolated hills. The layers of ash which formerly existed have been eroded away. These are common in central Scotland.

VOLCANIC LANDFORMS

Many spectacular landforms are also being created at present in the areas of the world that are experiencing active volcanic activity (Fig. 1.3). In contrast to the landforms so far discussed, which are primarily the product of weathering and erosion of the land surface, volcanic landforms are usually 'positive' in the sense that they have been built by volcanic activity, though, in some cases, explosive volcanic activity or subsidence has destroyed previously created landforms.

There are two major types of volcanic activity. In some cases lava erupts from a central crater. This tends to produce volcanic mountains that are roughly circular in plan. Many of these are liable to erupt violently but, in some cases, particularly if basalt lava is extruded, there may be relatively quiet eruptions. In other cases, lava flows relatively quietly out of fissures in the ground and may cover a large area to form a plateau. Volcanic landforms are created much more quickly than those that are the result of erosional processes.

Central volcanoes

One example of a type of central volcano is the cinder cone, in which ash and fragments of lava are extruded from a central vent. These can grow extremely quickly. Paricutin, in Mexico, first appeared as a cinder cone in 1943 and had grown to a height of 300 m in a few months. Monte Nuovo, near Naples, is said to have grown to 130 m in a week. They are not usually very high, rarely more than 300 m, and at first they suffer very little erosion because they are so porous.

Most well-known volcanoes are of the central type with alternating layers of ash, representing explosions, and lava, representing the emission of lava flows. Sometimes, parasitic cones may be built on the flanks of the volcano. They are sometimes called stratovolcanoes because of their layered structure. Explosions occur because the central vent becomes blocked by solidified lava, and pressure can build up for a number of years before an eruption takes place. Such volcanoes, therefore, erupt periodically after long periods of quiescence. In the case of Vesuvius, eruptions occur at intervals of several years. Stromboli, however, on an island off southern Italy, rarely explodes violently, but has very frequent eruptions of lava. The severity of the explosive eruptions is related to the viscosity of the magma. If the magma is very viscous, a severe eruption can take place, as at Mont Pelée in Martinique in 1902 when a spine of solidified lava rose 300 m above the crater and nuées ardentes (clouds of magma and gas) devastated surrounding areas.

Sometimes a devastating explosion or subsidence on a vast scale can destroy much of the structure of a central volcano and create a huge depression often bounded by steep walls, called a *caldera*. This happened in 1883 to the volcano Krakatoa, between Sumatra and Java, in Indonesia. The explosion is said to have been heard nearly 5000 km away. In the eruption of AD 79 much of the summit of Vesuvius was destroyed to form a depression called Monte Somma. Since then a new volcanic cone has been built, by successive eruptions, within the caldera of Monte Somma. Crater Lake, in the state of Oregon in the USA, occupies a nearly circular caldera over 9 km in width. This is believed to have been formed when an ancient volcano, Mt Mazama, subsided. Since then, a volcanic island (Wizard Island) has grown in the lake. Most of the world's calderas were once thought to have been formed by huge

The caldera of Las Cañadas in Tenerife

explosions but now it is believed that subsidence often played an important part.

Other central volcanoes contain magma of basic chemical composition and their eruptions are much less violent. Lava from these volcanoes can flow a considerable distance from the crater, at a fairly rapid speed, before solidifying. In an eruption in Iceland, lava is said to have flowed a distance of 60 km. Lava can also issue from fissures in the volcano's sides. Central volcanoes of this type have smooth, gentle (4–6 degrees) slopes which become flatter towards the summit. Because of their convex shape these volcanoes are called shield volcanoes. The best examples of this type are found in the Hawaiian islands in the Pacific and in Iceland where they are associated with the Mid-Atlantic Ridge. In the Hawaiian islands repeated eruptions have built up enormous structures rising some 9000 m from the bed of the Pacific, of which about 4000 m are above sea level. The flanks of these volcanoes are dissected by valleys where there have been no recent lava flows but, elsewhere, slope profiles are fairly smooth. New eruptions sometimes fill valleys with fresh lava. At the summit of Mauna Loa there is a broad, shallow caldera about 15 km in circumference. On its flanks is Kilauea with a lava lake in a smaller caldera. The volcanic activity of the Hawaiian islands is thought to be associated with a 'hot spot' on the floor of the Pacific Ocean. As the Pacific plate has moved north-west (Fig. 1.5) it has moved over this 'hot spot'. Thus, volcanic activity in the Hawaiian islands has moved towards the south-east. Volcanoes on the islands to the north-west of Hawaii are now extinct.

Lava plateaux

Lava plateaux are formed when lava wells up from fissures, often with very little explosive activity. The sheets of lava then cover the original relief. Numerous such eruptions can completely cover the pre-existing relief features and build up enormous plateaux. Such plateaux exist in the Deccan in India and the Drakensberg Mts in South Africa, both of which descend steeply to the coast of the Indian Ocean. They are also found in the Antrim Plateau in Northern Ireland and in the basins of the Columbia and Snake Rivers in the north-west of the USA. Here, the canyon cut by the Snake River is even deeper than the Grand Canyon. In lava plateaux the landscape is often shaped into steps composed of lava flows separated by weaker layers of volcanic ash. Parallel retreat of hillside slopes (Fig. 1.12) can produce steep-sided, table-shaped hills called 'buttes' (if small) and 'mesas' (if larger) (Fig. 1.38(*a*)).

THE INFLUENCE OF EARTH MOVEMENTS

Volcanic activity and earth movements, such as folding and faulting, can create differences in the level of the land surface but the relief features that they create have frequently been greatly modified by processes of erosion.

The influence of folding

Figure 1.38(*a*) need not represent the evolution of a lava plateau as suggested in the previous section. It could also represent a layer of horizontally bedded limestone resting upon a layer of weaker shale or clay. A large limestone plateau can be dissected by river erosion and the parallel retreat of valleyside slopes can dissect the plateau into smaller units (mesas and buttes). This sequence of events appears to have occurred very frequently in areas of the world with arid climates.

The folding of rock strata creates *anticlines* (upfolds) and *synclines* (downfolds), but it is comparatively rare for anticlines to form hill ridges and for synclines to form the intervening valleys. Such a relationship, known as 'normal relief' or 'primary fold relief', does exist in the Jura Mountains in eastern France on the fringe of the Alps, but there are relatively few examples elsewhere. It is much more common for anticlines to form relatively low ground, a relationship known as 'inverted relief'. The uplift of an area to form a structural dome does not mean that the highest ground will always be in the centre of the dome (Fig. 1.24(*b*)). After a period of erosion it may well be that the highest ground exists at the summits of the inward-facing escarpments. Inversion of relief can be brought about as follows. Earth movements first of all create a landscape of normal relief, in which the anticlines form hill ridges and the synclines form valleys, though no doubt some erosion would take place during this process. Streams drain from the anticlinal hills and join the trunk streams flowing along the synclinal valleys (Fig. 1.38(*b*)). These are all consequent streams (page 30). The streams draining down the flanks of the anticlines deepen their channels and may reach an underlying weak rock

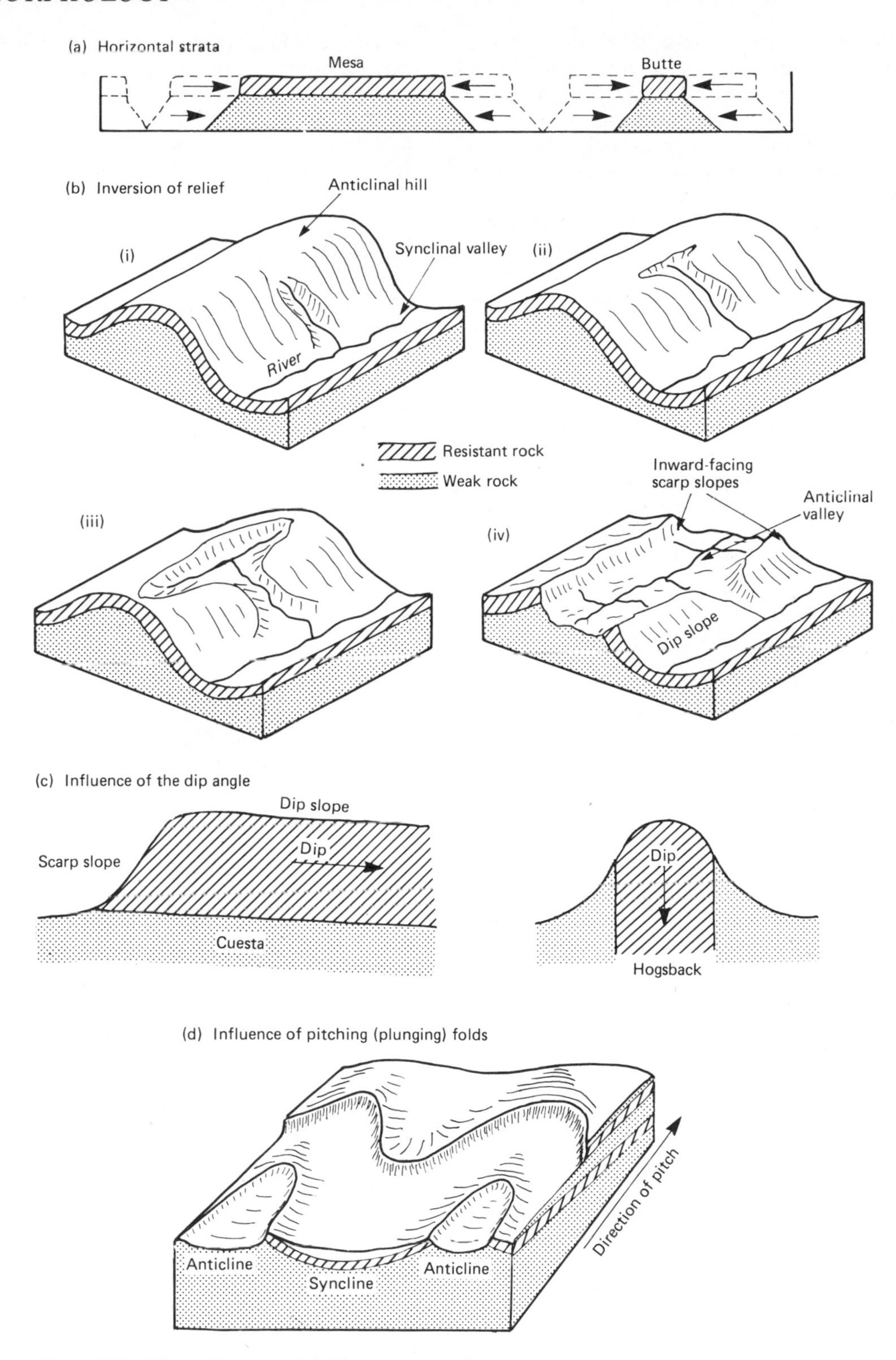

Fig. 1.38 The influence of folding on relief

layer. This encourages headward erosion by these streams and tributaries can develop in the weak rock that becomes exposed near the crest of the anticline. If several streams evolve in this way it is possible for river capture to occur (page 33) so that a long subsequent stream is formed on the weak rock along the axis of the anticline. Meanwhile, vertical erosion by the consequent stream flowing along the synclinal valley has been hindered by the presence of the resistant rock layer (Fig. 38(*b*)). Thus, an anticlinal valley has been created, with inward facing scarp slopes composed of the resistant rock layer.

Cuestas, hill ridges with a steep scarp slope on one side and a gentler slope on the other (Fig. 1.38(*c*)), commonly occur in areas where the strata have been folded. The gentler slope can only be referred to as a 'dip slope' if its slope is parallel to the dip of the strata. In other cases it should be called a back slope. The dip angle of the strata has a great influence on the profile of a cuesta. If the dip is very gentle, the scarp slope tends to be steep and the dip slope gentle. If the dip is very steep or vertical, a symmetrical hogsback tends to occur. Hogsbacks tend to have a lower elevation than cuestas composed of gently dipping strata since they are more susceptible to attack by weathering and erosion. Strong rocks, resistant to weathering and erosion, tend to produce steeper scarp slopes than weaker rocks, but it is possible for a scarp slope to be very steep if weaker rocks are overlain by a layer (cap rock) that resists erosion.

In some areas, pitching anticlines and synclines exist. Folds are said to pitch (or plunge) if the crests of the anticlines and the troughs of the synclines slope downwards longitudinally in one direction instead of remaining horizontal. In Figure 1.38(*d*) the crests of the two anticlines and the trough of the syncline slope downwards into the distance. This tends to results in the formation of sets of curving cuestas.

The influence of faulting

Faults are fractures in the earth's crust along which the rocks have been displaced either vertically or horizontally or at some intermediate angle. If faulting is to create relief features, movement along faults must take place at a faster rate than erosional processes are able to destroy the developing relief features.

The fault-plane along which movement takes place can be either vertical (Figs. 1.39(*a*) and (*b*)) or inclined (Figs. 1.39(*c*) and (*d*)). In *normal faults* the rocks have been displaced in the direction towards which the fault-plane is inclined (Fig. 1.39(*c*)). In *reverse faults* they have been displaced downwards in the opposite direction (Fig. 1.39(*d*)). In the case of a *tear fault* displacement is horizontal (Fig. 1.39(*g*)).

A common landscape feature that results from faulting is the *fault-scarp*. Ideally this is the fault-plane itself, but usually its form has been much altered by processes of erosion. In some cases the original fault-scarp may have retreated parallel to itself so that it no longer marks the actual location of the fault. In other cases it may have been dissected by stream erosion so that it now exists as triangular facets between which deep valleys have been excavated (Fig. 1.39(*b*)).

In the case of a reverse fault the actual fault-plane may not be visible at the surface because of slumping as the higher block rose over the lower block (Fig. 1.39(*d*)).

If a relatively narrow block of rock subsides between two normal faults (Fig. 1.39(*e*)), a rift valley (*graben*) is formed. It is possible, however, that some rift valleys have been formed by parallel reverse faults (Fig. 1.39 (*d*)) and that the slumping observed on the valley sides gives a false impression of subsidence of the valley itself. This may be the case where rift valleys cross relatively high ground. The great rift valleys of east Africa extend for over 3000 km from north to south and they are occupied by long, narrow and very deep lakes, such as Lake Tanganyika and Lake Malawi. This system of rift valleys extends northwards along the Red Sea and continues in the Dead Sea and the Jordan valley, where its floor is well below sea level. This great system of rift valleys has been interpreted by some as the first stages of the opening-up of a new ocean. A remarkable example of a rift valley is the Turfan depression in central Asia, whose floor descends to below sea level although it is located thousands of kilometres from the coastline. Another good example of a rift valley is that occupied by the River Rhine to the north of Basel. This rift valley is flanked by two horsts (Fig. 1.39(*f*)), the Vosges and the Black Forest. Block faulting, in which the land surface is shaped into alternating horsts and rift valleys, is characteristic of large areas in the plateaux west of the Rocky Mountains in the USA, particularly in the state of Utah. In the far west the Sierra Nevada is a huge faulted block.

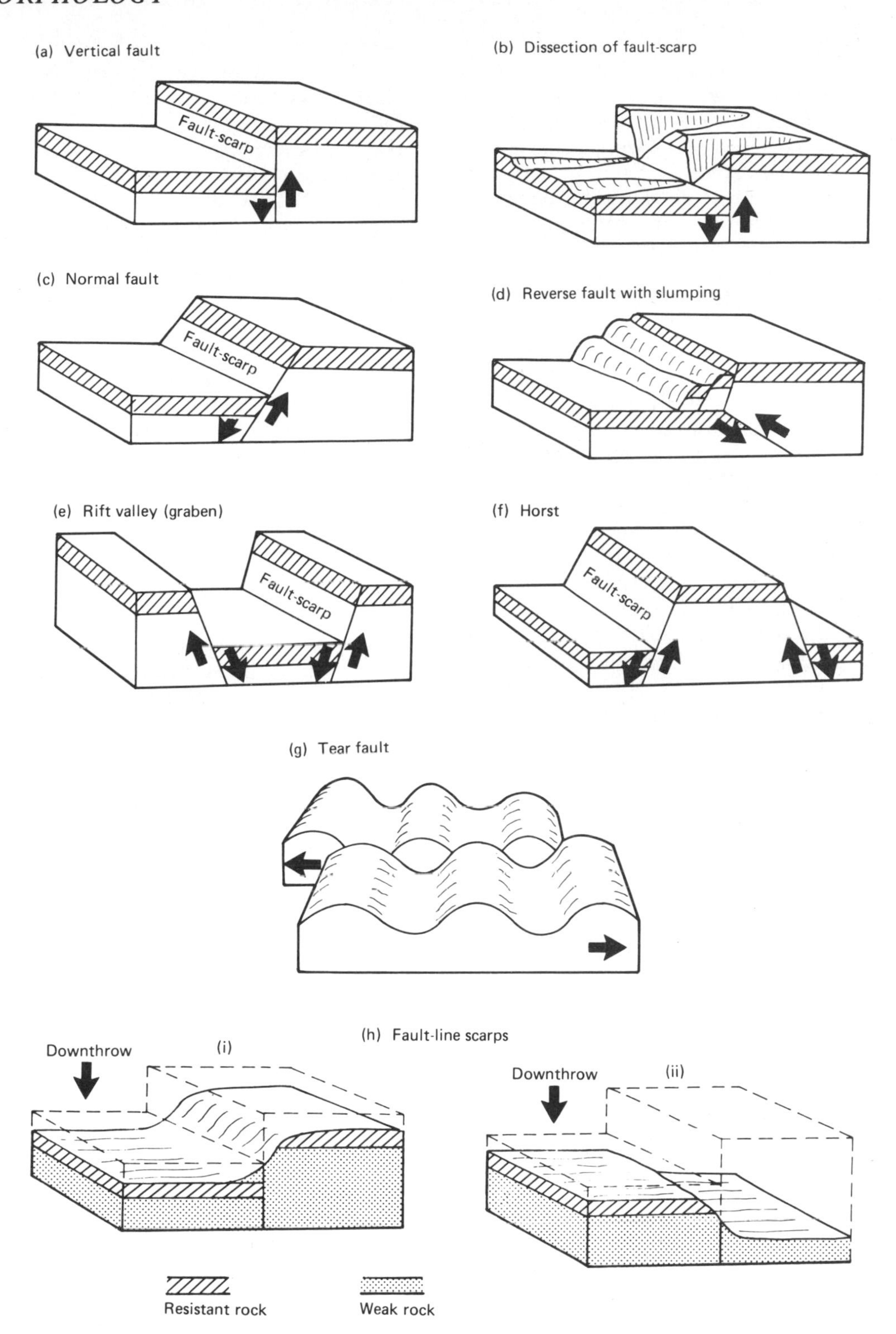

Fig. 1.39 The influence of faulting on relief

Figure 1.39(*g*) illustrates the characteristics of tear faulting. A common effect of this type of faulting is that valleys are blocked by ridges which have shifted laterally. Streams therefore tend to change their courses quite suddenly at the line of the fault. The San Andreas Fault, which runs northwards from the Gulf of California, is an example of tear faulting. This fault marks the eastern margin of the Pacific crustal plate which is moving northwards and thus forming a zone of crustal instability at its junction with the North American plate (Fig. 1.5).

The relief features of a fault-line scarp, in contrast to those of a true fault-scarp, are the result of erosion. The form of the scarp owes more to the variations in the resistance of the rock layers to erosion than it does to the actual process of faulting. In Figure 1.39(*h*)(*i*), the downthrow of the fault is towards the left and the original fault-scarp faced towards the left. Subsequently some erosion has taken place. The fault-scarp has become somewhat degraded, but the scarp still faces in the original direction, and it does so because of the position of the layer of resistant rock. In Figure 1.39(*h*)(*ii*), the downthrow of the fault is towards the left, but the scarp faces in the opposite direction. This is because the resistant layer has protected the area to the left of the fault from erosion. Genuine fault-scarps which are the direct result of faulting are found mostly in areas that are tectonically active at the present time, where earthquake activity is common (Fig. 1.3). In other areas, such as the United Kingdom, fault-line scarps are more common. No large-scale faulting has occurred in the United Kingdom in recent geological times.

1.5 Desert landforms

Landforms in deserts have a very different appearance from those in humid environments. This is because of the special characteristics of the desert environment. Deserts have long periods without any rain at all. Hence, the ground surface is usually dry and can be attacked by wind erosion. When rain does fall it tends to be evaporated very quickly, so there is less surface run-off than elsewhere, but a great deal of run-off may occur for a short time as a result of heavy convectional storms. Much evidence exists of former wetter periods in deserts that have particularly affected their poleward margins. River erosion during these periods may have created the wadis (dry valleys) that now exist in the northern Sahara for example. Since there is so little vegetation in deserts there is little interception of rainfall, so much of the ground surface is exposed to erosion. During heavy rain the ground's infiltration capacity may soon be exceeded, thus producing considerable overland flow quite suddenly. Showers capable of generating overland flow are most frequent on the margins of the deserts. The regolith (soil) in deserts tends to be quite shallow, and bare rock frequently exists at the ground surface. This encourages overland flow and causes desert landforms to be more angular in shape than those found in more humid areas.

CONTRASTS IN DESERT LANDSCAPES

There is no single set of landforms that can be regarded as characteristic of all desert areas. Great contrasts exist between the different deserts of the world. This is not surprising. Arid and semi-arid areas extend from north-east Asia to the equator in East Africa, and their climates differ a great deal. The mean temperature of the coolest month varies from under 0°C to over 20°C and that of the warmest month from under 20°C to over 30°C. Some deserts, such as the central part of the Sahara, are very dry, but some of the desert margins can receive considerably more rain. The Atacama Desert in western South America receives very little rain, but it is affected by mists which provide a humid atmosphere that can encourage chemical weathering processes. The season of maximum rainfall also varies, even within a single area of desert. On their equatorward side, for example, the hot deserts, such as the Sahara, receive their rain mostly during the hot season, when evaporation is particularly great. Hence there is unlikely to be very much overland flow. On their poleward side they have a winter rainfall maximum and less is lost by evaporation.

Drainage conditions also vary within a single area of desert. Some rivers rise within the desert, so their flow is intermittent and only occurs after a period of rainfall. Other rivers, such as the Nile, the Colorado and the Indus, rise outside the desert and flow across it. They can therefore maintain their flow perennially across the whole width of the desert, though evaporation may reduce their discharge considerably. They are thus able to create 'normal' fluvial landforms in an area that has a desert climate. These rivers are able to transport their load of sediment to the sea and can therefore be influenced by changes in base level (page 45). Other rivers flow into enclosed basins where they may form lakes from which water either percolates into the ground or is evaporated. Such rivers are unaffected by changes of base level. Other desert areas may only have intermittent streams.

Desert areas also differ in their geological structure. Some, such as the Sahara and the Australian deserts, have been affected relatively little by earth movements and have large areas of subdued relief. These are located on major crustal plates (Fig. 1.3). On the other hand, the deserts of south-west USA and the Pacific coast of South America are located at plate margins, and their landforms owe a great deal to earth movements.

Sand dunes are comparatively rare in deserts. It has been estimated that less than 20% of the world's deserts have a sand-covered land surface, and sand dunes are particularly rare in south-west USA. Much larger areas have a land surface of bare rock or gravel. In some deserts, mountains may occupy almost half of the total area, as in the Sahara and Arabia. In south-west USA the land surface has been divided by faulting into horsts and grabens (Fig. 1.39) and alluvial fans have been built up along the fault scarps. In the centres of the downfaulted basins, salt marshes (playas) are found, sometimes covered by shallow lakes after rain. In places, permanent lakes exist, such as the Great Salt Lake of Utah. Drainage lines converge on the centres of such basins except in the case of large, permanent rivers such as the Colorado. In very dry deserts exposed rock surfaces have been little affected by weathering, so the detailed geological structures are clearly visible. In wetter areas there is greater coverage of the ground surface by weathered and eroded rock debris.

WEATHERING IN DESERTS

Physical weathering appears to be more effective than chemical weathering in deserts (pages 7 and 8) but it is likely that much of the weathering that takes place is the result of a combination of physical and chemical processes.

Desert climates are characterized by a very large diurnal range of temperature, up to 20°C to 30°C, and this temperature variation is even greater on rock surfaces that are sunlit during the day than in the atmosphere. Rocks are generally poor conductors of heat, so a steep temperature gradient develops. Hence, the surface layers of rocks suffer stresses caused by alternate expansion and contraction. This can cause the surface layer to fracture parallel to the original surface, a process called exfoliation. The effect of these sudden temperature changes seems to be greater if the rock surface is occasionally wetted by rain, dew or mist. Exfoliation tends to produce a mass of coarse, angular rock fragments. Another type of physical weathering is granular disintegration (page 7) whereby the rock mass breaks up into its constituent mineral grains. The mineral grains that make up rocks react differently to changes of temperature because they have different colours and also different coefficients of expansion. Sudden changes of temperature therefore set up stresses within the rock mass and cause it to disintegrate into fine-grained particles. Also, dust containing various salts may be carried in the wind and then deposited by rain on rock surfaces. These salts can develop crystals which can prise apart the particles of solid rock.

Chemical weathering is now believed to be more important in deserts than was once thought. Chemical weathering processes require the presence of water and deserts receive very little precipitation. But considerable amounts of moisture can exist for a time on desert surfaces as a result of the formation of dew and mists. The high diurnal range of temperature tends to produce dew and some coastal deserts in particular tend to have frequent mists. Also, chemical weathering tends to increase in importance towards the desert margins where there is more moisture and also more vegetation to produce organic acids. On the shaded side of upstanding rock masses, rounded hollows a few metres across sometimes occur, especially in deserts that are liable to mists. Such hollows are known as 'tafoni' and they are believed to have been formed by hydration (page 8) of mineral particles which causes them to expand and to set up stresses that disintegrate the solid rock.

EROSION AND DEPOSITION BY WATER

Water is probably the main agent of erosion in

deserts. Desert climates are characterized by occasional heavy convectional rains. In these heavy showers precipitation can exceed the infiltration capacity of the land surface and overland flow can occur. The lack of vegetation means that there is little interception of precipitation. It is also likely that present-day desert areas had more rain and less evaporation during the 'pluvial' periods of the Ice Age, which occurred when the ice sheets advanced. Many desert streams flow into permanent or temporary lakes or salt marshes or playas. Here, ancient shorelines, indicating former, higher lake levels, can frequently be seen.

Wadis and streamflow

Stream channels in deserts are usually dry but, when rain occurs, vigorous streamflow can take place along the wadis (dry valleys). This usually only affects a small part of any network of wadis. These temporary streams rapidly lose water by evaporation and seepage into the ground. The floors of the wadis usually contain a great deal of rock debris. Hence, as the stream flows along, its load increases rapidly but the volume of its discharge decreases. Hence, the stream may become a mudflow and forward motion may cease (Fig. 1.40 (*a*)), though some erosion can take place in the steeper parts of the profile before the stream becomes fully loaded. The typical stream hydrograph shows a rapid rise in discharge, with little lag between the occurrence of precipitation and peak discharge. All discharge is quickflow (Fig. 1.17). Also, in a downstream direction, peak discharge decreases very quickly (Fig. 1.40 (*b*)). The stream channel itself can take the form of a network of minor channels, constantly joining and dividing, rather like a braided stream (page 39). The origin of the wadis, along which streamflow takes place, is in some doubt. In some cases it is possible that they have been excavated by present-day stream erosion, but usually it is more likely that they were formed in a period when the deserts received a greater rainfall. Major systems of wadis are probably too large and complex to have been created by streamflow in present-day desert conditions. Also, some wadis have steep gradients at their upper ends which suggests spring sapping (page 50) at a former water table. Rivers that flow perennially across the desert, usually with sources outside the desert, produce the usual landforms of fluvial erosion and deposition.

Slopes, pediments and inselbergs

The action of water is also important on the flanks of desert uplands or on the margins of downfaulted depressions (grabens) that are surrounded by fault scarps. Here, slope elements of varying steepness tend to be separated by comparatively

(a) Streamflow along a wadi

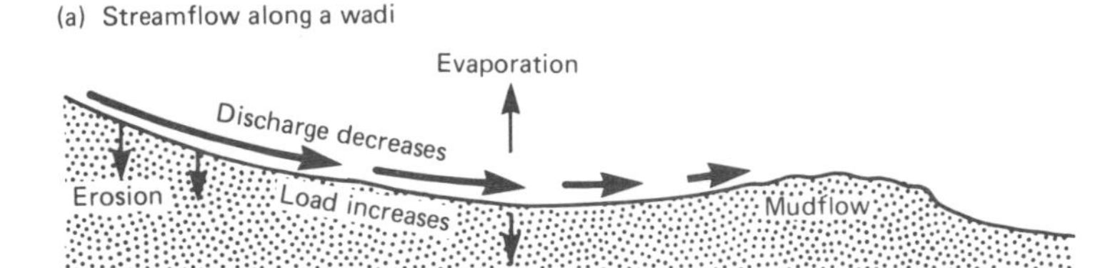

(b) Hydrographs at four locations spaced at intervals along a desert stream channel

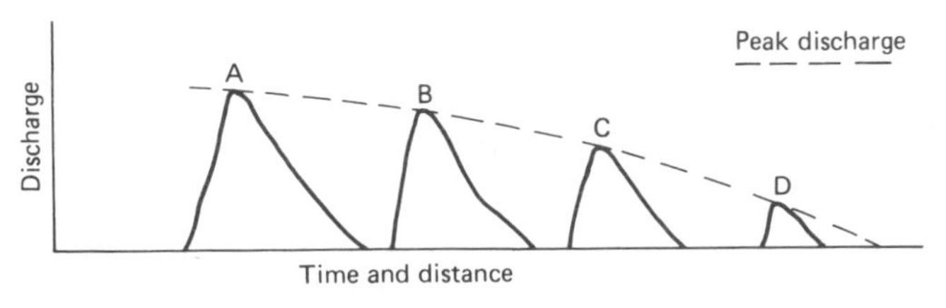

Fig. 1.40 Streamflow in a desert

sharp breaks of slope instead of the smoothly curving concavo-convex profiles that are so common in humid climates (Fig. 1.41).

The mountain front or escarpment is often very steep and may sometimes be vertical. If the rocks are of a uniform type this slope tends to be rectilinear (straight in profile) (Fig. 1.41(*a*)). Otherwise its slope tends to reflect the differing resistance to weathering of its various rock layers. The lower ground in front of the escarpment is a gently sloping solid rock surface known as a pediment. The escarpment often meets the pediment at a very sharp break of slope (a *knick*), a feature that is only found in semi-arid areas. This

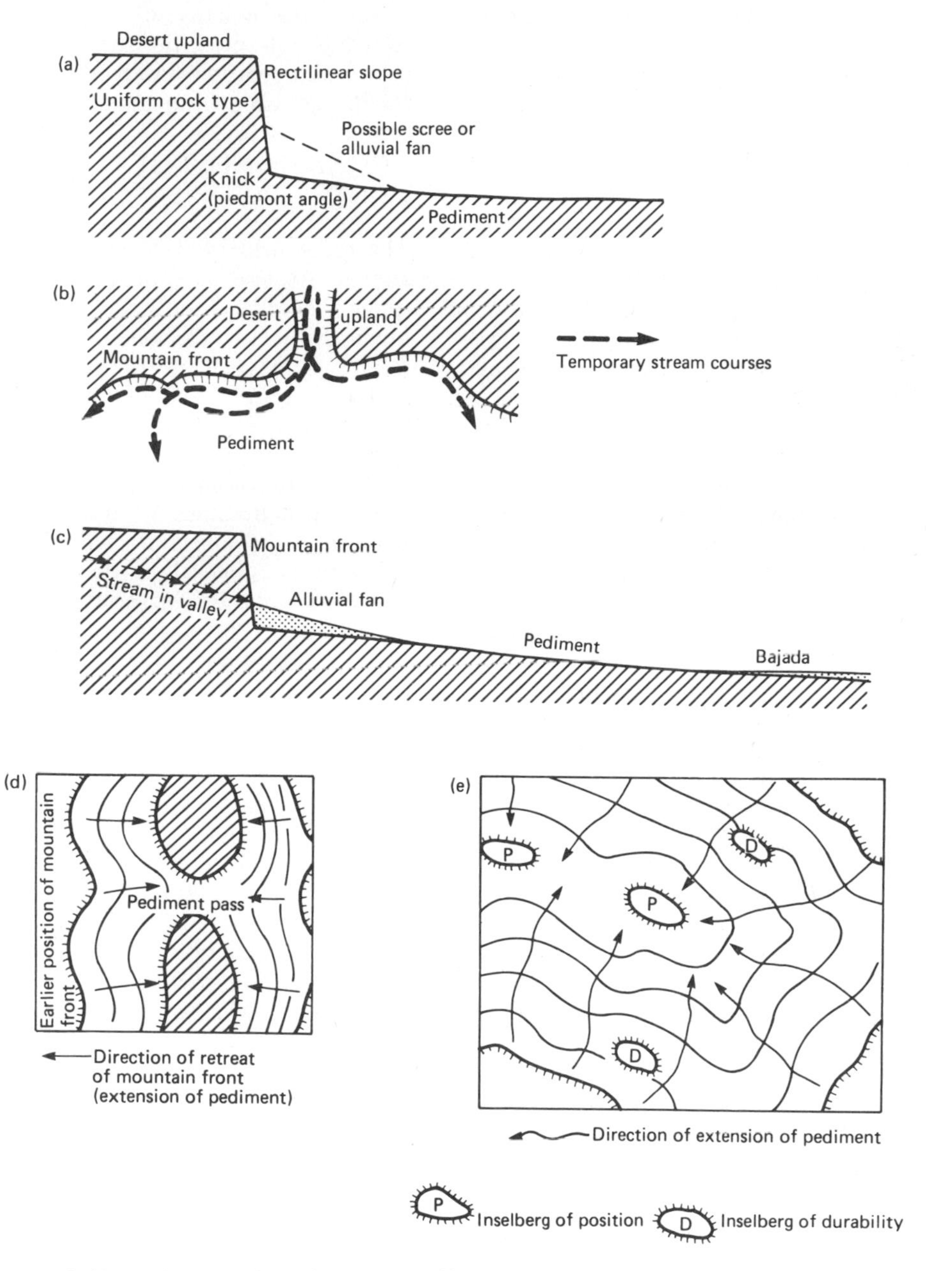

Fig. 1.41 Pediments, bajadas and inselbergs

knick is sometimes called the piedmont angle (*piedmont* being a term used to describe the upland margin in general) (Fig. 1.41(*a*)). The profile of the pediment is slightly concave upwards, becoming less steep further from the mountain front. The pediment may be slightly dissected by shallow drainage channels which are usually dry.

Little is known about the origin of the knick. Some believe that it could be the result of lateral erosion by streams issuing from the upland area and undercutting the base of the mountain front. In this case, however, one would expect the mountain front to be curved in plan (Fig. 1.41(*b*)). This is not always the case. Knicks can be seen where the mountain front is perfectly straight. Another possible explanation is that water collects at the base of the escarpment and, as a result, weathering is particularly active there. In places the piedmont angle may not be visible because screes have accumulated at the foot of the escarpment (Fig. 1.41(*a*)) or large alluvial fans have been built by streams issuing from the upland area.

The pediment itself functions as a surface over which weathered debris from the mountain front is transported to a lower level by moving sheets of water called sheet floods which occur after heavy rain. Sheet floods are able to transport medium-sized pebbles. Sediment thus transported across the pediment accumulates at a lower level as a series of gently sloping alluvial fans, known as *bajadas*, which tend to encroach upon the lower edge of the pediment. The term 'bajada' is also used to refer to the rather steeper alluvial fans that spread from the mountain front over the upper edge of the pediment (Fig. 1.41(*c*)).

The mountain front is believed to retreat parallel to itself (Fig. 1.12). As it does so the pediment gradually advances into the upland area. It may advance particularly rapidly into upland valleys and may eventually cut straight across an upland area to form a pediment pass (Fig. 1.41(*d*)). Parallel retreat of the mountain front and accompanying expansion of pediments can begin at a fault-scarp or at the steep sides of wadis. Eventually, upland areas can be destroyed and reduced to isolated upstanding rock masses called inselbergs. Two major types of inselberg are identified according to the reason for their existence. 'Inselbergs of position' owe their existence to their location at a considerable distance from the depressions where pediment formation first began. They are the last remnants of a former desert upland. 'Inselbergs of durability' owe their existence to their rock type, which has successfully resisted weathering and erosion. This type of inselberg can exist very near to where pediment extension began (Fig. 1.41(*e*)). Inselbergs have many varied shapes. Exfoliation weathering can produce rounded inselbergs. If they are strongly jointed they may be more angular.

Since the expansion of pediments by slope retreat begins in a variety of locations a region can have pediments at a variety of different levels, with inselbergs rising from each level.

EROSION AND DEPOSITION BY WIND

The movement of wind over a desert surface is similar to that of water over a stream bed (page 37). The wind's velocity increases in an upward direction because, near the ground surface, it is reduced by friction, particularly if the surface is rough. The wind's power to transport particles of sediment varies according to their size. Very fine particles (dust) can be lifted and carried in suspension, sometimes to great heights. A dust haze can sometimes be observed from aircraft flying at 10 000 m near the Sahara. Sand grains are fairly easily rolled along the ground or they can move by saltation (a series of short 'hops') in which the sand grain is raised almost vertically and then falls gradually to the ground. It is rare for sand grains to be lifted higher than about two metres.

Erosion by the wind

The wind is capable of performing two kinds of erosional processes. *Abrasion* is erosion of the ground surface by the impact of moving sand particles. *Deflation* is the lifting and removal from the ground surface of loose particles.

In the process of *saltation* the sand grains may perform erosion in two different ways. When they fall back to the ground, at a considerable speed, they may disturb other grains and set them in motion. Also, they may be driven against any upstanding rock masses. Since saltation can only take place very near ground level it is possible that the undercutting of vertical rock faces and the development of mushroom-shaped pedestal rocks are the result of wind abrasion. However, it is also possible that these landforms are the result of the operation of weathering processes near ground level and the removal of weathered particles by deflation. Wind abrasion cannot take place at higher levels because only dust particles can be lifted to any considerable height and particles as fine as dust are carried by the wind round any obstacle without actually striking it. Hence, there are relatively few landforms that can definitely be attributed to abrasion by the wind. It seems likely, however, that yardangs have been shaped by wind abrasion. *Yardangs* are irregularly shaped ridges extending parallel to the prevailing wind. They reach a height of about 10 m and are separated by U-shaped troughs. They are composed of rather weak sediments and their upper surfaces and sides are fluted in a series of smoothly concave hollows, separated by sharp angles.

Much more impressive landforms have been attributed to the wind's ability to excavate *closed hollows* (deflation hollows) by lifting and carrying away loose sand which may have been produced by the weathering of a relatively weak rock. The deepening of such a hollow appears to be able to continue until either the water table is exposed or until the hollow provides enough shelter to reduce the wind's strength. Many of these hollows are relatively small, only a few metres deep, but the Qattara depression in Egypt, with an area of about 50 000 sq. km and extending to over 100 m below sea level, has been attributed to wind deflation.

Deposition by the wind

The wind transports sand and then, under certain conditions, deposits it to create a variety of landforms. Landforms of wind deposition are most common in or near the great 'sand seas' (*ergs*) where the sand probably originated as alluvial deposits in pluvial stages of the Ice Age.

Some sand accumulations are static; others are mobile. Sand drifts tend to form in the shelter of some obstacle such as an upstanding rock or a cliff, where the wind's velocity is suddenly decreased. They can also form at the mouth of a gorge where the air stream diverges and slows down. The size of such sand drifts is generally determined by the amount of shelter that exists, combined with the fact that the maximum angle of rest of dry sand is about 34 degrees (Fig. 1.42(*a*)). Sand dunes, on the other hand, do not depend for their existence on shelter. They can move forward with the wind and their shape is directly influenced by the wind's direction and strength.

There tends to exist a hierarchy of wind-deposited landforms. Sand ripples, a few centimetres high, are the smallest group. The next larger group consists of dunes, such as barchans and seifs, whose surfaces may be covered with sand ripples. Finally, dunes can exist on the surfaces of huge sand ridges, known as *draa*, which may be at least 300 m high. The *barchan* is a crescent-shaped dune that is concave in the direction towards which the wind is blowing and has a pair of 'horns' that project downwind (Fig. 1.42(*b*)). Barchans assume their most perfect crescent shape when the wind tends to blow from a constant direction and on a compact, stony surface where there is comparatively little sand. The sand that makes up a barchan may be derived from a sand drift that has accumulated behind an obstacle, and several barchans may be arranged in a line downwind of such a drift. A barchan's windward slope is quite gradual, but its leeward slope (slip face) can be as steep as 34 degrees. Barchans are often about 400 m wide and about 30 m high, though they can be much smaller. The 'horns' tend to move forward faster than the centre, but they move partly into the shelter of the main mass. Thus a crescent shape is established that can only be disturbed by a change in the direction of the wind. Sand is driven up the windward slope and is deposited near the top of the slip face thus steepening it. When the slope of the slip face exceeds 34 degrees, slumping takes place (Fig. 1.42(*b*)) and the dune moves forward. Small dunes tend to move forward faster than large dunes. Barchans are comparatively rare since very few areas have winds from a constant direction.

In some cases, *parabolic dunes* are formed. These are curved in the opposite direction to barchans, being convex in the direction towards which the wind is blowing, with wings extending upwind. The curve of the dune partially encloses a deflation hollow (Fig. 1.42(*c*)).

Seifs are long dune ridges extending parallel to the direction of the dominant wind. Some believe that they can be formed by the coalescence of a number of barchans whose shape has been distorted by cross winds. They can be over 100 m high and at least 100 km long. Several parallel, evenly spaced seifs can occur, with barchans in the intervening troughs where they are sheltered from cross winds. Slip faces on the summits and slopes of seifs are evidence of cross winds at an angle from the dominant wind.

In areas where sand is abundant it is common to

find a combination of barchans and parabolic dunes linked together to form winding ridges running generally at right-angles to the dominant wind. These are believed to have been formed by eddying in the air within the general flow of the wind. The parts of the ridges that resemble barchans are located where the airflow is comparatively slow and tends to converge in a downwind direction. The 'parabolic' sections (convex in a downwind direction) appear to be shaped by a fast divergent flow of wind. Moving downwind the barchan type and the parabolic type occur alternately (Fig. 1.42(*d*)). In some cases the oblique sections of these dunes may coalesce to form seifs.

1.6 Landforms of glaciation and periglaciation

The most recent major period of glaciation occurred in Pleistocene times and came to an end between 10 000 and 15 000 years ago. In this short period of time the landforms created by this glaciation have changed their appearance comparatively little. In the Pleistocene Ice Age there were four major periods during which glaciers advanced to cover large areas of North America,

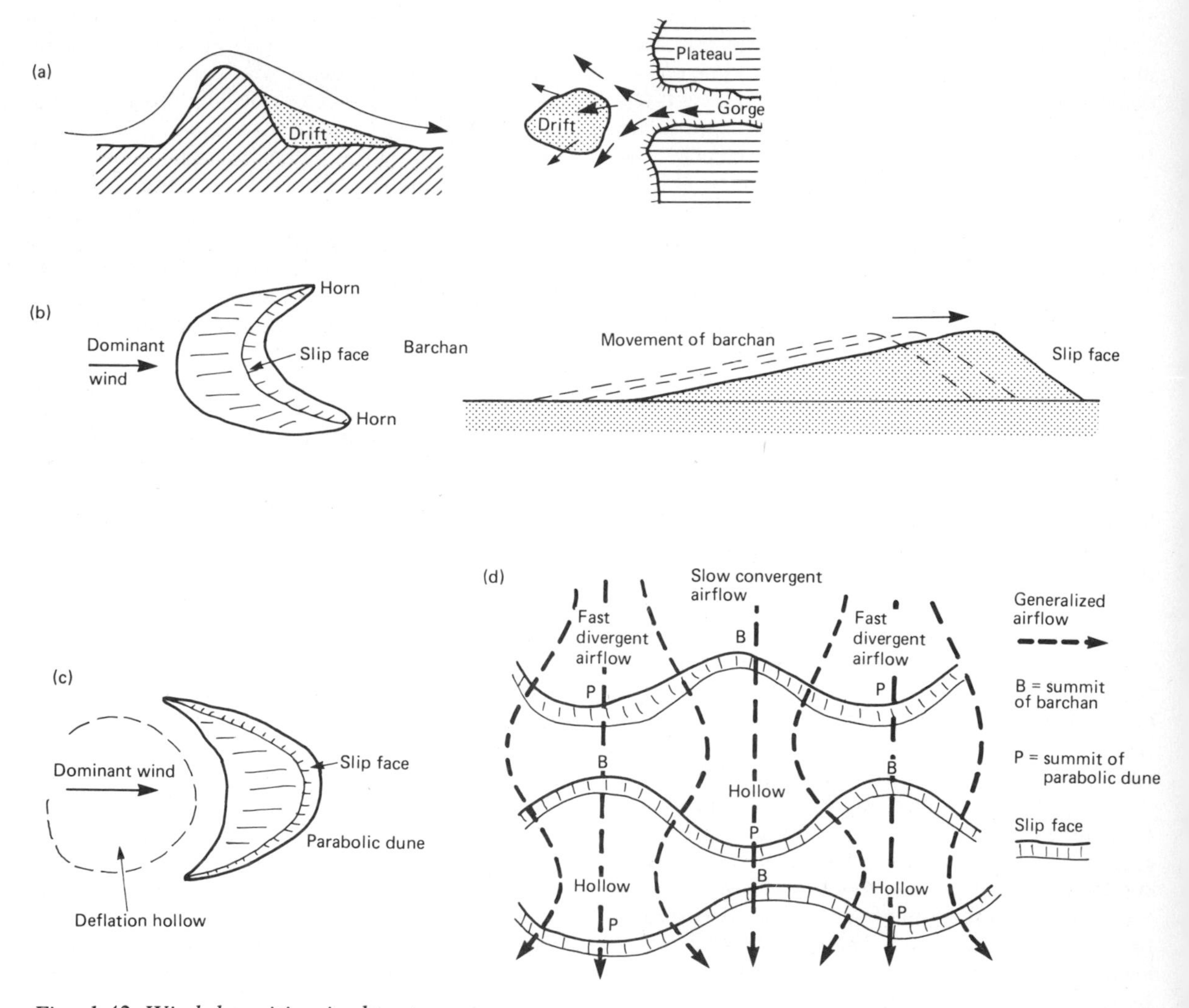

Fig. 1.42 Wind deposition in deserts

Europe and Asia. These four periods are termed the Gunz, Mindel, Riss and Würm glacial stages. At each stage approximately the same area was covered by ice. Between these glacial stages there were much longer periods (interglacial stages) when the ice retreated and the climate was sometimes warmer than at the present time. It has been estimated that, during a glacial stage, temperatures were up to 10°C cooler than present-day temperatures, and that ice occupied an area about three times as large. Today, ice sheets, ice caps and glaciers cover about 10 % of the world's land area. Antarctica makes up over 80 % of this area and Greenland makes up over 10 %. Considerable areas of ice also exist in high latitude islands such as Iceland and Spitzbergen and also in north-east Canada and Alaska. Much smaller ice-covered areas exist in temperate latitudes. These are situated in mountainous areas where low temperatures are caused by the high altitude. The main glaciated areas of this type are the Himalayas and other mountain ranges in Asia, the Alps, western Canada and New Zealand. During the twentieth century the world's glaciers have generally tended to retreat.

CHARACTERISTICS OF GLACIERS

The accumulation of ice

Ice can accumulate in an area and form a glacier if the average amount of snow that falls in winter is greater than the amount that can be melted or evaporated (*ablation*) in summer. Also, in some cases, extra snow can be delivered to a developing glacier by avalanches from nearby steep slopes. A glacier is created by the transformation of snow into glacier ice. Newly fallen snow has a very low density because of the large amount of air that it contains between its crystals. Pressure from overlying snow can change the snow crystals into coarse grains. Then, after about a year (including a summer melting season), the granular snow is changed into a denser substance known as *firn*. Finally, after a period of several years, the firn, probably through pressure from overlying snow, becomes glacier ice. Ice tends to accumulate most rapidly on the leeward sides of hills and in deep valleys that are shaded from the sun.

Types of glaciers

The largest ice formations in the world are the huge ice sheets of Antarctica, which are more than 4000 m thick, and Greenland. During the Ice Age much of Britain was covered by an ice cap centred on highland areas such as the Scottish Highlands, the Lake District and Wales. This ice cap sometimes expanded to meet the Scandinavian ice sheet in the area of the North Sea. The movement of ice sheets and ice caps is sometimes influenced by the relief of the underlying ground surface, but it is not controlled by it. Much movement of ice takes place in response to the thickness of the ice. Ice tends to flow in the direction in which the ice surface slopes (Fig. 1.43(*a*)). This means that ice can flow uphill over the ground surface and can therefore transport rocks from a lower to a higher level.

Valley glaciers are streams of ice that are strongly controlled by relief features. They begin in hollows (cirques) at the upper ends of valleys, where snow collects and is changed into glacier ice. They then flow downhill along a pre-existing valley. Such glaciers are common in mountain ranges such as the Alps and the Himalayas. Sometimes a single valley glacier may have several tributaries each of which starts in a cirque (Fig. 1.43(*b*)). Valley glaciers tend to achieve a state of dynamic equilibrium in which the rate of accumulation of snow and ice at their upper ends is balanced by the rate of wastage by ablation at a lower height. If the climate becomes warmer the rate of melting may increase and this may cause the ice front to retreat. A reduction in snowfall can have a similar effect (Fig. 1.43(*c*)). Similarly, an increase in snowfall or a cooler climate can result in a lengthening of the glacier. Valley glaciers descend considerably below the snow line. This is necessary in order to achieve a rate of melting that matches the rate of accumulation. A different kind of valley glacier is one that flows downhill from an ice cap that occupies an area of high ground. Such glaciers exist in Iceland, Greenland and Norway (Fig. 1.43(*d*)).

In some cases, a valley glacier may advance beyond the confines of its valley and spread out to form a large sheet on an adjacent plain. This formation is called a piedmont glacier (Fig. 1.43(*e*)). A well-known example is the Malaspina glacier on the coast of Alaska which reaches a thickness of 600 m and is thought in some places to flow slightly uphill over the ground surface.

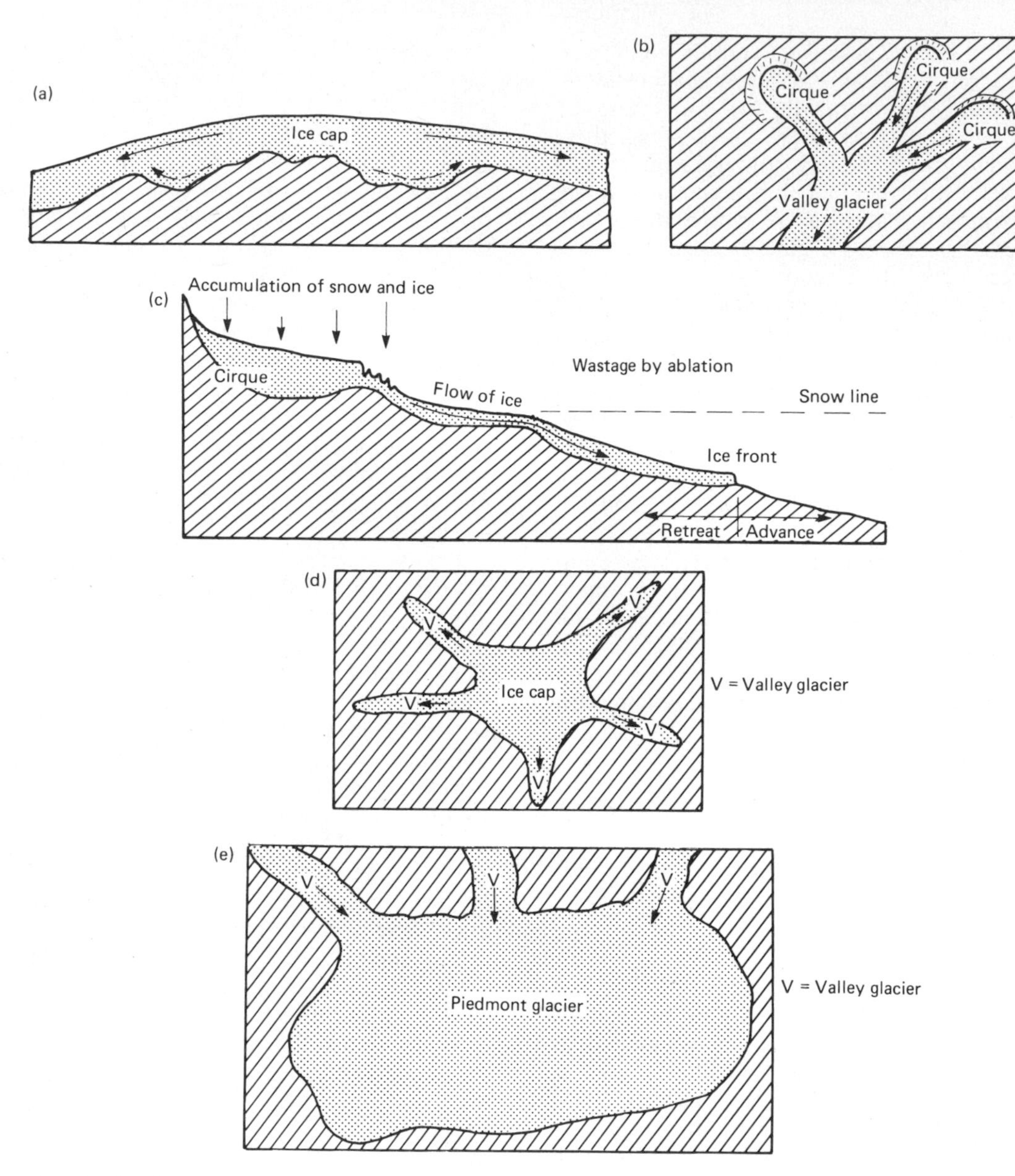

Fig. 1.43 Glaciers

EROSION AND DEPOSITION BY ICE

During the Ice Age glaciers and ice sheets created a wide variety of landforms by erosion particularly in highland areas. With the retreat of the ice, these features have been revealed. The processes of erosion by ice have been investigated by studies of existing glaciers. As the ice sheets retreated they deposited the rock debris derived from erosion, sometimes by direct deposition from melting ice and sometimes by deposition from meltwater streams. This section is concerned with the work of ice. The work of meltwater will be considered later.

Erosion by ice

The major processes involved in erosion by ice are abrasion and plucking or quarrying. *Abrasion* is the erosion of the underlying rock surface by fragments of rock held in the base of the glacier. This process is thought to be particularly effective in the case of valley glaciers in the mountainous parts of temperate regions, such as the Alps where the base of the glacier can slide fairly easily over the underlying surface. Some think that in parts of a glacier where the ice is relatively thick, because of a depression in the valley floor, the ice near the base of the glacier can flow relatively quickly (extrusion flow) and erode its bed particularly effectively, thus deepening the depression. *Plucking* or *quarrying* involves the lifting and removal of joint-bounded blocks of rock from the bed of the glacier. Such blocks may be loosened by the freezing and expansion of meltwater in the joints that surround them or by the dragging effect of the base of the glacier as it passes over them. The plucking process may also be assisted by the release of pressure on the underlying bedrock as the glaciated valley is deepened.

Landforms created by glacial erosion

At the upper end of a glaciated valley there is usually a *cirque* (or cwm or corrie). This is a large rounded depression, with a steep back wall, sometimes containing a lake which may occupy a rock basin or may be dammed by moraine. The creation of a cirque can begin by snow accumulating in a pre-glacial hollow round about the level of the snow line. In the northern hemisphere slopes facing between north and east are most favourable because here melting is less in the summer months. The original hollow is then enlarged by *nivation*: in the summer and by day the snow melts and percolates into the underlying rock, which it disintegrates by frost shattering in cold weather (page 8). The rock debris is then removed by *solifluction*. Eventually the hollow becomes large enough for snow to collect and be transformed into firn and eventually into glacier ice (page 65). In time, a cirque glacier comes into existence and spills out of the hollow to feed a valley glacier. Since the ice flows from the cirque glacier to the valley glacier that is at a lower level it is not clear why the cirque has a rock lip at its exit (Fig. 1.44(*a*)). This can be explained by the manner in which the cirque glacier moves. In winter the upper part of the cirque glacier, near the back wall, receives more snow than the lower part. In summer, however, melting is greater in the lower part than the upper part. Hence, the glacier's surface becomes steeper (Fig. 1.44(*a*)). Eventually the glacier rotates and, in so doing, tends to deepen its basin. Another problem concerning a cirque is that the back wall appears to retreat over its whole height by frost shattering even though its lower part appears to be protected by the glacier from temperature variations. It has been suggested that meltwater could penetrate down the bergschrund crevasse, between the glacier and the back wall, and attack the lower part of the back wall by freeze-thaw, but no final solution has been found. As cirques develop, their back walls and sides tend to retreat and they become larger and larger (Fig. 1.44(*b*)). If two neighbouring cirques meet, an arête, a sharp-topped, jagged ridge, is formed. If three or more cirques meet, a pyramidal peak can be formed. Sometimes cirques can occur in succession along a glaciated valley, separated by steep 'steps', as in the case of Glaslyn and Llyn Llydaw on the flanks of Snowdon. Some high mountains that have been glaciated have no cirques because their rocks are too weak to support steep back walls.

The long profile of a glaciated valley is usually irregular, with rock basins and steep rock steps, especially at higher levels. The rivers that now occupy such valleys have irregular long profiles with frequent waterfalls, rapids and lakes. A rock step is often found near the head of the valley where several cirque glaciers converged to form the valley glacier (Fig. 1.44(*c*)). Other rock steps appear to be associated with a narrowing of the valley that might have increased the thickness of the glacier. Rock basins are sometimes excavated where strongly jointed rocks occur next to massive, coherent rocks in the valley floor, or below a pre-glacial river's knickpoint (page 45). One theory suggests that if the ice flow at the glacier's surface is decelerating as the glacier passes over a depression in its bed, shear planes in the ice curve upwards and rock debris can be carried to the glacier's surface (Fig. 1.44(*d*)). This can help to deepen the depression. *Roches moutonnées* are common features on the valley floor. These have a smoothly rounded, striated *stoss side* (facing up-valley). The lee side is steeper and more irregular (Fig. 1.44(*e*)). Here, large joint blocks have often been plucked out by the glacier. In glaciated valleys, plucking often seems to have had a greater effect than abrasion. The view looking up a glaciated valley reveals many more irregular

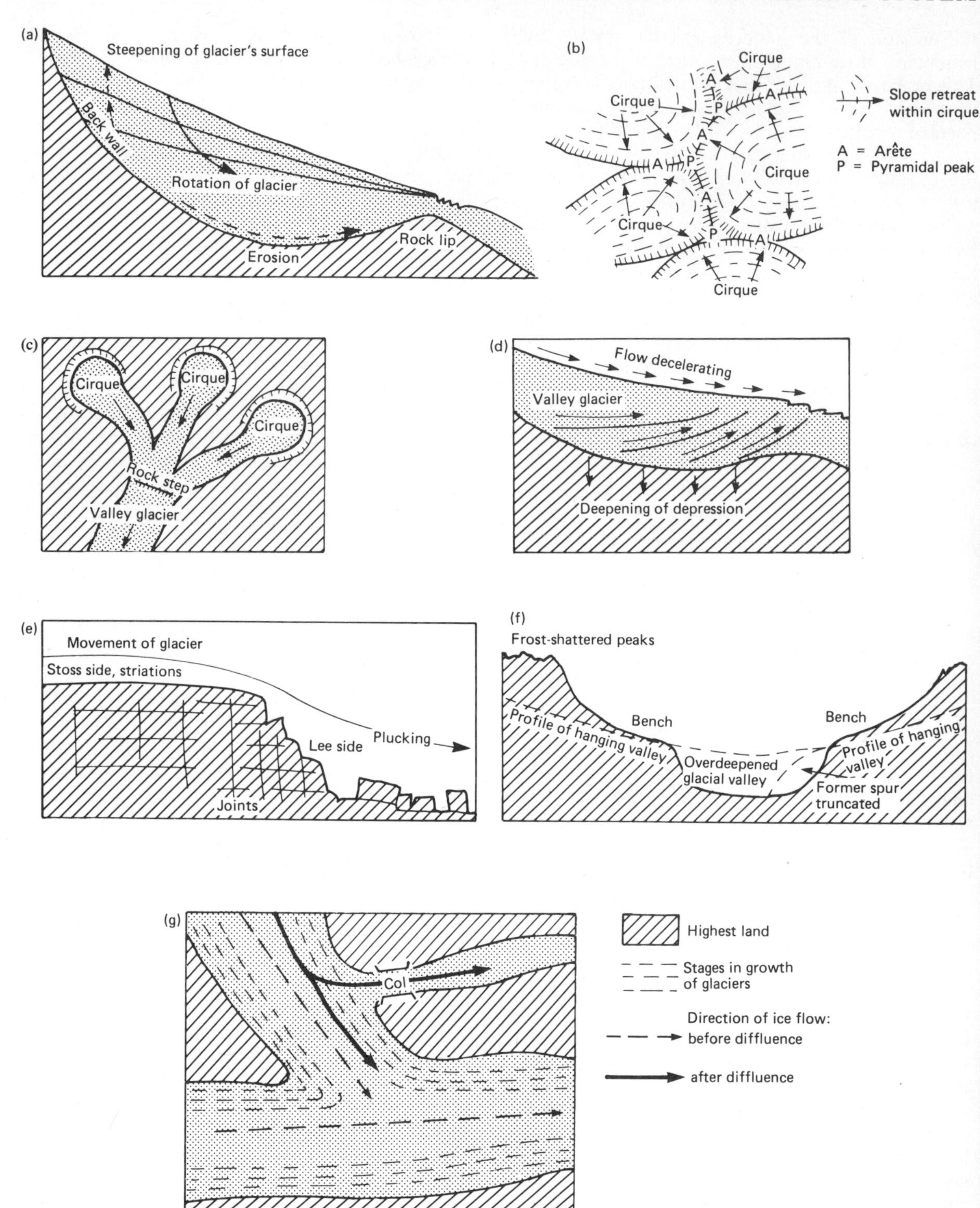

Fig. 1.44 Landforms of glacial erosion

A glaciated valley

craggy features than the view looking down the same valley. When a glaciated valley reaches the sea coast it usually continues seawards as a fjord. *Fjords* are particularly common where glaciated highlands reach the coast, as in Alaska, Norway and north-west Scotland. They can be very deep indeed. The Sogne Fjord in Norway is over 1000 m deep. At their seaward end there is a 'threshold' where the water suddenly shallows. Because of these thresholds it appears that fjords are mainly the result of glacial erosion. The depth of water in fjords is much greater than the post-glacial rise of sea level, so it appears that glaciers performed erosion at a depth well below the sea level of the time. Theoretically this should be possible. When ice floats, nine-tenths of its volume is below water level. Hence, one would expect a glacier 1000 m thick to be able to erode its bed beneath the sea to a depth of about 900 m.

A glaciated valley's cross profile is often described as U-shaped, the sides being very steep, sometimes almost vertical, and the floor being generally flatter and wider than a river-eroded valley. Spurs that formerly extended from the valley sides have been truncated by erosion. Relatively level benches (alps) have been left high up on the valley sides and tributary valleys often descend steeply to the main valley floor (Fig. 1.44(*f*)). All these erosional features could have been created by either vertical or lateral erosion by the glacier. If vertical erosion were the main process one would expect the projected profiles of the hanging valleys to meet above the level of the valley floor (Fig. 1.44(*f*)). In some cases the valley's cross profile has been affected by glacial diffluence. This occurs because the glacier's outlet at the lower end of its valley becomes blocked, possibly by another glacier. Ice therefore builds up in the valley and the glacier becomes thicker and thicker. Eventually the upper part of the glacier overflows through the lowest *col* (saddle) on the valley side and flows into a neighbouring valley (Fig. 1.44(*g*)). The level of the col can then be reduced by erosion and erosional landforms may be created in the col.

In glaciated lowland areas that were once covered by ice sheets or ice caps depositional landforms are usually more common than those created by ice erosion. Here, the ice flowed much more slowly than in mountainous areas. Nevertheless, in areas such as the Canadian Shield and lowland areas in north-west Scotland, rounded rocky hills and rock-basin lakes frequently occur. The rock basins have commonly been excavated along zones of weakness created by outcrops of relatively weak rocks, faults or closely spaced joints. Some relatively low hilly areas, such as the Pennines, were completely covered by ice at the maximum advance of the ice sheets, so the erosional landforms described above are usually absent. The influence of rock type is also important. Weak rocks such as shales cannot be shaped into the striking landforms which exist in areas such as Snowdonia, the Lake District and the Scottish Highlands.

Landforms created by glacial deposition

Deposition by an ice sheet often leaves a layer of glacial deposits up to 300 m deep covering the pre-glacial relief and forming a level or gently undulating plain (a *till plain*). Streams that originated on this drift cover have courses that are quite unrelated to their pre-glacial courses and bear little relationship to the relief of the underlying solid

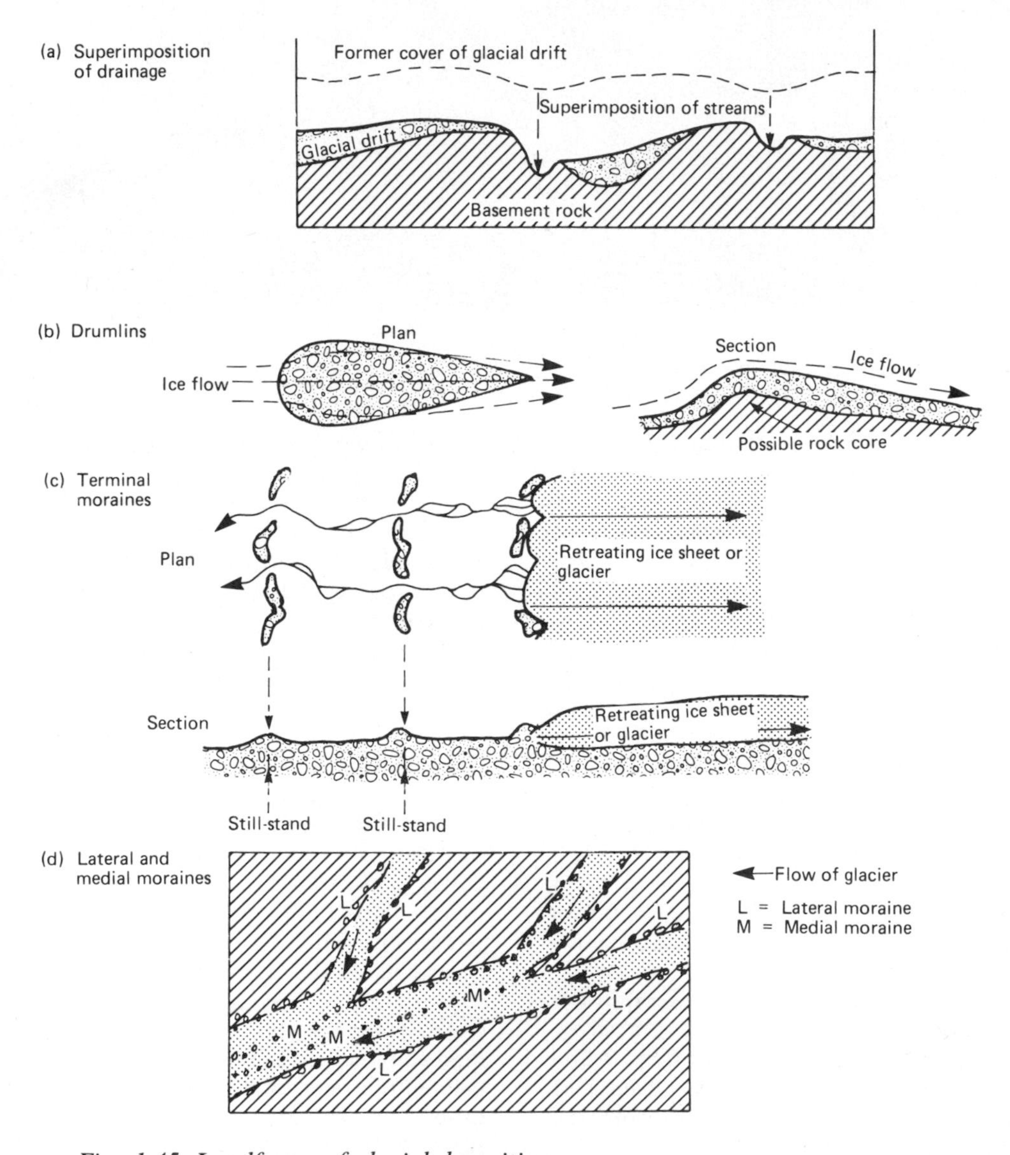

Fig. 1.45 Landforms of glacial deposition

rock. Thus, gorges or river cliffs may be excavated in the basement rock (Fig. 1.45(*a*)). This is a type of superimposed drainage (page 34) under rather unusual conditions. The glacial deposits themselves often consist largely of boulder clay but they also include sand and gravel. It is often possible to identify rock fragments which have been transported a great distance from their source area, thus making it possible to reconstruct the direction of movement of the ice sheet. Such rocks are called 'erratics'. Many granite erratics carried by ice from the island of Ailsa Craig, in the Firth of Clyde, have been found in north-west England. Low streamlined hills known as 'drumlins' occur, particularly in areas that were affected by the most recent advance of the ice sheets. These may be anything from 6 to 60 m in height. They tend to be regularly spaced over the landscape and they are elongated in the direction in which the ice sheet moved. The length of a drumlin is usually between two and four times its breadth. A greater length than this suggests that the ice sheet flowed particularly quickly. The steeper, broader end of a drumlin faces in the direction from which the ice sheet came; the opposite end is narrower and more gently sloping (Fig. 1.45(*b*)). Their origin is not completely understood but it is believed that they represent accumulations of moraine under the ice sheet that were shaped by the movement of cleaner ice on each side. Some drumlins have a rock core that may have impeded the movement of the ice and resulted in deposition. Drumlins are particularly common in the north of Ireland and on each side of the Solway Firth.

Terminal moraines can be built at the outermost margin of an ice sheet or at the lower end of a valley glacier. They are usually created at a time when the ice front is stationary. This means that the amount of ice brought forward along the glacier can just be melted at the ice front. Thus, the glacier's load is deposited at the ice front. A renewed forward movement of the glacier would destroy the terminal moraine. If the rate of ice flow along the glacier decreases, or the rate of melting increases, the ice front retreats to a new equilibrium position. The last retreat of an ice sheet or glacier is often marked by a series of terminal moraines, each having been built at a stillstand of the ice front. These are called recessional moraines (Fig. 1.45(*c*)). Ideally a terminal moraine forms a curving ridge running parallel to the ice front, but often valleys have been cut through it by meltwater streams and it can sometimes comprise a number of separate hillocks.

Lateral moraines are built up along the valley sides by scree deposits, avalanches or possibly rock fragments removed from valleyside slopes by ice erosion. If two glaciers meet, their lateral moraines can merge and form a *medial moraine* that runs along the length of the glacier below the junction (Fig. 1.45(*d*)). Medial moraines rarely survive as distinct landforms after glaciation, but occasionally they can be seen as a short continuation of the spur between two glaciated valleys at a junction of glaciers.

EROSION AND DEPOSITION BY MELTWATER

As the ice sheets and glaciers melted at the end of a period of glaciation a great deal of meltwater was produced that resulted in the creation of large numbers of temporary streams. These streams produced a variety of landforms by both erosion and deposition.

Landforms created by fluvioglacial erosion

As a period of glaciation came to an end it appears that in many areas glaciers and ice sheets still existed in the larger valleys and the lowland areas as meltwater flowed out of smaller valleys. Thus, the flow of meltwater could have been blocked by barriers of ice and temporary lakes could form. Further melting could raise the level of such lakes so that they overflowed across watersheds. The resulting erosion could create new valleys or gorges passing across the watershed. Such valleys are referred to as overflow channels or spillways. Several different types of fluvioglacial spillways have been identified (Fig. 1.46). A direct overflow is one in which meltwater flowed towards the head of the tributary valley and away from the edge of the ice (Fig. 1.46(*a*)). At the present time the River Derwent in North Yorkshire has its source near the coast, near Scarborough. It then flows inland and passes through a gorge at Kirkham Abbey before joining the Ouse and reaching the sea. It has been suggested that a large lake covered the Vale of Pickering, dammed at its eastern end by North Sea ice. The lake overflowed in the south-west (a direct overflow) thus creating the Kirkham Abbey gorge.

In other cases the lower ends of a number of valleys may have been blocked by an ice barrier,

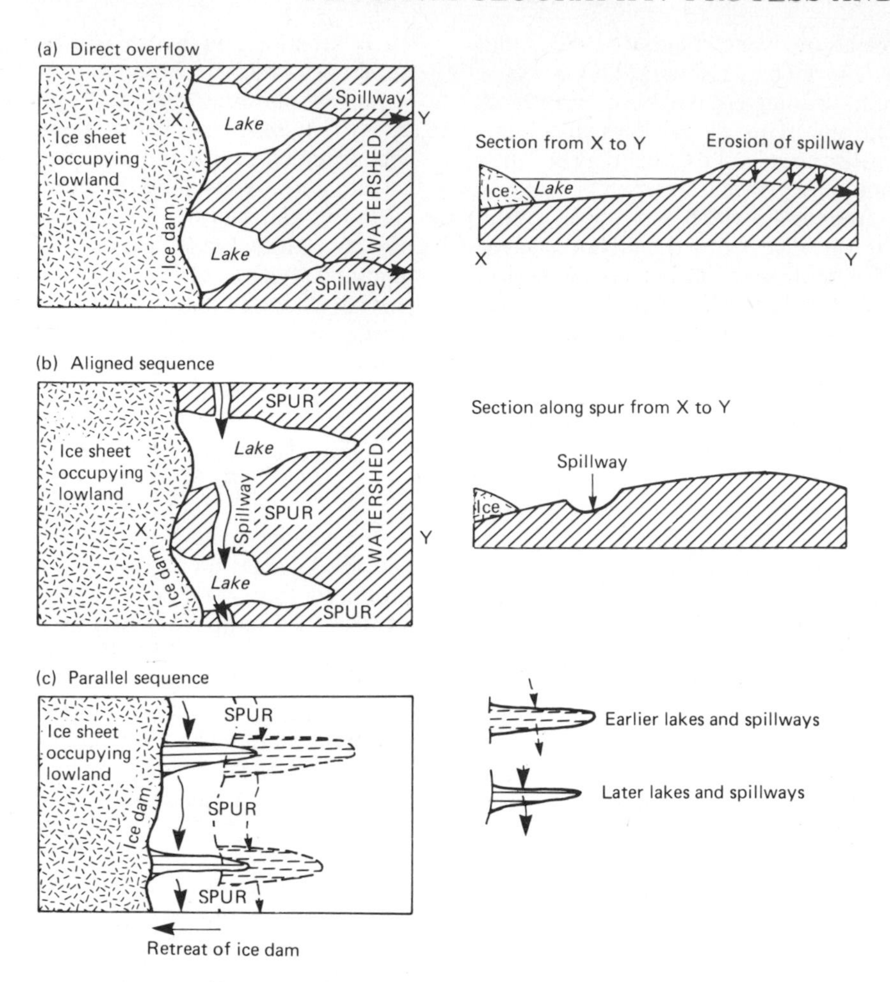

Fig. 1.46 Landforms of fluvioglacial erosion

lakes formed, and these lakes overflowed from valley to valley producing notches in the successive spurs (Fig. 1.46(*b*)). In these cases the spillways are likely to be dry at the present time. Such a system is called an aligned sequence. Sometimes, in such a sequence, the ice barrier would slowly waste away, thus gradually uncovering successively lower parts of the spurs, so that the lake overflowed at successively lower points along the spur. This is referred to as a parallel sequence (Fig. 1.46(*c*)).

The result of successful research into such spillways has been that notches that existed in the profiles of spurs probably tended to be too readily interpreted as fluvioglacial spillways.

Other possibilities do of course exist. Every notch in the profile of a hillside or a spur need not necessarily have been created by water overflowing from a pro-glacial lake. Some could easily have been carved by streams flowing along the junction between the glacier and the hill slope. Also, erosion of the underlying ground can be performed by streams flowing under hydrostatic pressure within and under the ice. Such streams can also flow uphill and can therefore cut channels across spurs.

Landforms created by fluvioglacial deposition

There are two general types of landforms that have been created by fluvioglacial deposition. Proglacial deposits are those that have accumulated as a result of meltwater deposition beyond the margin of the glacier or ice sheet. On the other hand, ice-contact landforms have been created by meltwater streams flowing over or within the ice or on its margins.

The most important pro-glacial landform is the *outwash plain* which is usually a series of alluvial fans. Streams, heavily loaded with sands and gravel, flow from the ice front, which may be either the lower end of a valley glacier or the margin of an ice sheet. These streams have braided courses (page 39) which frequently change their position. The streams tend to deposit the coarser gravels nearest to the ice margin and the finer sands further away. Outwash plains build up to their greatest size when the ice front is stationary. The surface of an outwash plain is often pitted with small depressions known as *kettle holes*. These have been formed by the melting of blocks of ice which have been covered by fluvioglacial material. When the ice melts the fluvioglacial cover collapses and leaves a small depression which may become a small lake (Fig. 1.47(*a*)). The retreat of a valley glacier has sometimes resulted in the accumulation of outwash deposits fairly evenly over a considerable length of the valley floor. This feature is referred to as a *valley train* (Fig. 1.47(*b*)). In some cases it gives the valley floor a very smooth surface which contrasts with the eroded valley sides and emphasises the U-shape of the cross profile. In other cases river erosion has

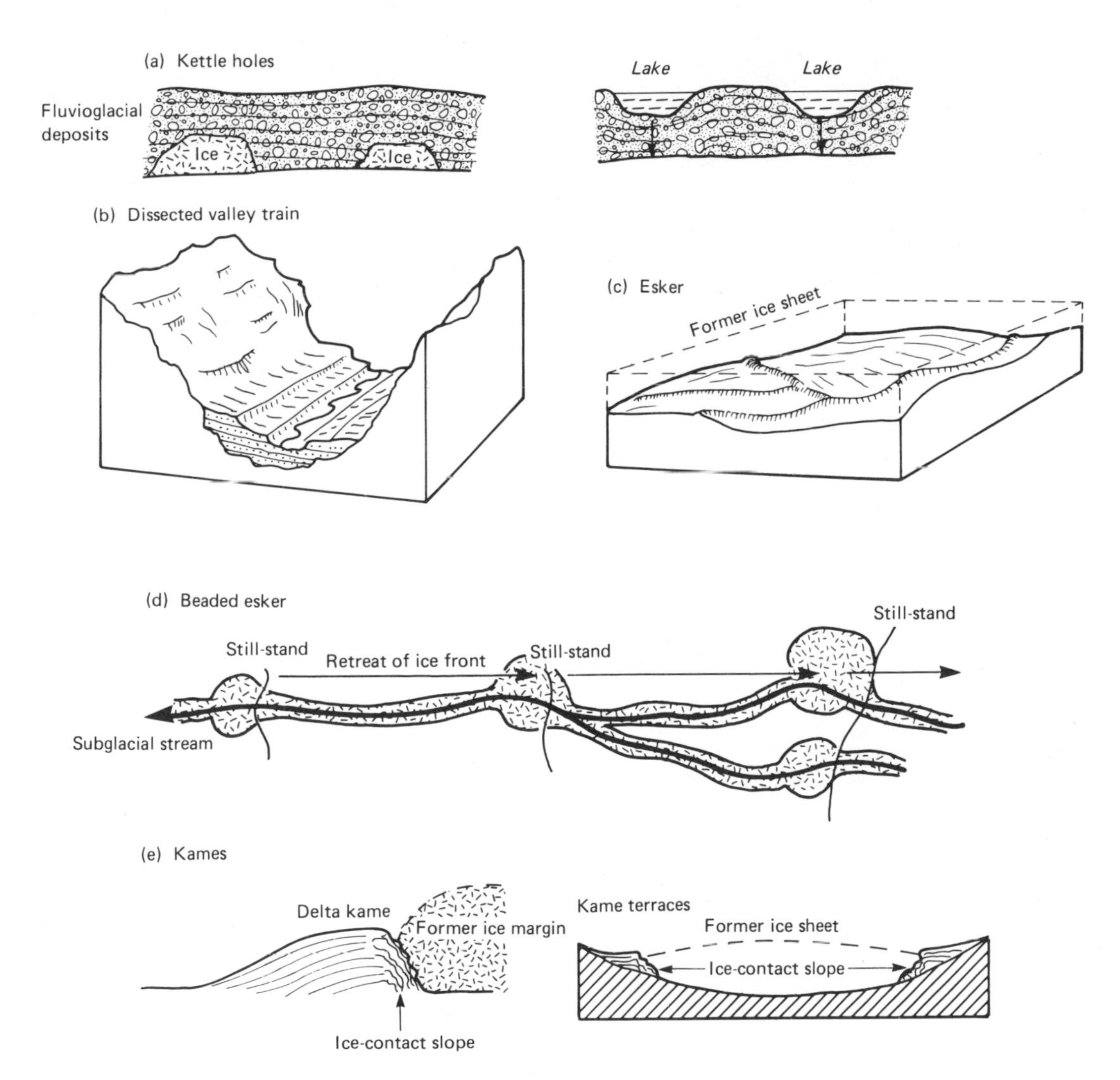

Fig. 1.47 Landforms of fluvioglacial deposition

Drumlins in south-west Scotland

The internal structure of an esker

dissected the valley train, thus creating river terraces (page 46 and Fig. 1.47(*b*)).

The main fluvioglacial, ice-contact landforms are eskers and kames. An *esker* is a long (sometimes reaching 100 km), winding ridge composed of fluvioglacial sands and gravels. Such ridges are sometimes joined to other similar ridges forming a pattern similar to that of a river and its tributaries. Eskers do not only occur on low ground. It is common for them to cross low ridges which is extremely unusual for landforms created by deposition by water. Some eskers may have been formed by deposition in tunnels beneath or within a stagnant, decaying ice sheet. Streams flowing under or within the ice sheet, and under pressure, can carry large loads of sediment. If their exit near the ice margin becomes blocked by sediment, deposition can occur along their whole course. When the ice sheet finally melts the former course of the subglacial or englacial stream is represented by a ridge of sediment that has been superimposed on the subglacial surface (Fig. 1.47(*c*)). Alternatively an esker could be formed by a retreat of the ice front with continuous deposition by a subglacial stream gradually forming a ridge. If the retreat of the ice front is occasionally interrupted by stillstands, a beaded esker can result (Fig. 1.47(*d*)). More sediment is deposited at the ice margin during a stillstand than when the ice front is retreating. Eskers are associated more with the decay of stagnant ice sheets than with valley glaciers.

Kames are of several different types. Some (delta kames) are created by the building of deltas in proglacial lakes near the ice front. The stratification of their sands and gravels indicates their manner of growth. Delta kames may exist as isolated low mounds, with an irregular slope on the side that was formerly in contact with the ice margin (ice-contact slope) (Fig. 1.47(*e*)). Crevasse kames are isolated low ridges formed by the infilling of the ice sheet's crevasses with fluvioglacial material. Such deposits can trend in any direction according to the former pattern of the ice sheet's crevasses. Kame terraces form relatively level terraces along the sides of a valley that was formerly occupied by a stagnant ice sheet. Sediment was deposited in the trough between the ice sheet and the valley side. When the ice sheet melted the deposits resting on its edges collapsed to give a fairly steep, irregular ice-contact slope (Fig. 1.47(*e*)). Other fluvioglacial deposits were created as beach deposits, sometimes associated with wave-cut benches on the shores of glacier-dammed lakes which now no longer exist. Three such benches, sometimes with rounded water-worn pebbles, exist at different heights in Glen Roy, near Fort William in the Scottish Highlands. They are known as the Parallel Roads of Glenroy. They represent three different levels of the shoreline of a lake that existed in Glen Roy during the Ice Age.

PERIGLACIAL PROCESSES AND LANDFORMS

Periglacial processes are those that are active in cold climates but are not associated with the presence of glaciers. They are related to the alternate freezing and thawing of the ground. In the relatively warm summers the surface layers of the ground thaw but, at a greater depth, the ground is permanently frozen, a condition that is referred to as permafrost. But periglacial mass wasting processes can operate where permafrost is absent. Thus, it is difficult to define the term 'periglacial' precisely. Periglacial processes are associated with a variety of environments. They occur predominantly in the cold tundra and forest areas of high latitudes, but they also occur in lower latitudes, at high altitude, in mountain ranges. During the Ice Age, periglacial processes were active on the equatorward fringes of the ice sheets. Thus, evidence of former periglaciation exists today in many temperate areas.

Landforms created by frost action

Periglacial areas are particularly affected by the alternate freezing and thawing of the regolith. By these processes the mantle of rock waste is shaped into various characteristic forms.

When moist sediments freeze they first expand as ice begins to form. Then, as the temperature falls further, they begin to contract. This tends to form a set of cracks. The following summer, meltwater fills the cracks and, in turn, freezes, thus widening the cracks. Eventually, thick wedges of ice develop which exert pressure on the surrounding sediments to dome them upwards into a series of low mounds (Fig. 1.48(*a*)). The cracks themselves eventually become filled with sediment as the climate becomes warmer. Such 'fossil' ice wedges are often found in present-day temperate regions and are evidence of former periglaciation.

The development of ice wedges could thus lead

to the creation of a series of domes and depressions on the ground surface, known as patterned ground (Fig. 1.48 (*b*)). This consists of domes and depressions of the ground surface regularly arranged in patterns of circles, polygons or stripes, according to the slope of the ground. On very gentle slopes circles and regular polygons tend to be formed. As the slope steepens these are elongated in the direction of the slope until eventually they form stripes (Fig. 1.48 (*b*)). In some cases the patterned ground may be 'unsorted' and may consist entirely of fine-grained sediment, the troughs sometimes being marked by lines of vegetation. In other cases it may be 'sorted', the domes of finer sediment being surrounded by troughs containing larger stones. In these cases it appears that the larger stones have been heaved upwards by frost action and have then accumulated in the troughs.

On a much larger scale is a dome-shaped hill, known as a *pingo*, which may be up to 60 m in height and several hundred metres in diameter. A pingo is formed by water or saturated soil becoming enclosed by frozen sediment. This may occur where a lake protects the underlying sediments from freezing. The lake becomes infilled with sediment, which freezes from the top downwards. Also, the surrounding permafrost encroaches from the sides (Fig. 1.48(*c*)). Eventually the mass of saturated, unfrozen lake sediments becomes enclosed. As this gradually freezes, it expands upwards, where resistance is weakest. This forms a dome (a pingo). This may eventually collapse to form a rounded hollow, surrounded by a rampart. This type is known as a closed-system pingo. An open-system pingo can be formed if, for example, artesian water invades the waste mantle and then freezes.

Landforms created by mass wasting processes

The mass movement of weathered sediments tends to form large-scale accumulations of debris in periglacial areas. It is common to find masses of fairly large, angular blocks of rock on, or at the foot of, steep slopes. Such coarse debris is produced by the shattering of bedrock by alternate freeze and thaw operating along joints (page 8). Debris falls freely from the steeper slopes or is transported down the slopes by avalanches or frost creep, which involves the lifting of sediment at right-angles to the ground surface during periods of frost and vertical subsidence when thawing takes place (Fig. 1.48(*d*)).

Accumulations of coarse debris take the form of screes at the foot of free faces, and large areas of angular blocks known as *blockfields* or *felsenmeer*. Blockfields may have been produced by rock glaciers which are able to flow down slopes because of the presence of ice in their lower parts. In some cases, however, the blocks have been carried down slopes by mudflows, and the mud has later been washed out by rain.

Finer-grained debris can be transported down slopes by various processes. The term slushflow is used to describe the transport of fine-grained debris along valley floors by large amounts of water derived from the melting of snow in summer. This debris is deposited in unsorted masses, forming irregular fans at the mouths of valleys. *Slushflow* is intermediate between mass movement and streamflow. A much slower process is *gelifluction*, the special case of solifluction in areas of frozen ground. In summer, the surface of the waste sheet melts while the underlying layers remain frozen. Thus, a sheet of debris is able to flow down very gentle slopes. Periglacial mass movement of frost-shattered chalk during the Ice Age in Britain has given rise to the rounded coombes that exist on the scarp slopes of chalk uplands such as the South Downs (Fig. 1.48(*e*)). In front of such coombes, on the lower ground, large fans of periglacial rock debris can frequently be seen. Also, many of the chalk dry valleys have accumulations of poorly sorted, periglacial rock debris, known as *head*, along their floors.

Landforms created by nivation

Nivation is the term used to describe the processes of weathering and mass wasting that take place beneath and at the edges of snowdrifts. Meltwater from the snowdrift percolates into the underlying rock and then freezes, thus tending to shatter the rock. The shattered debris is then removed by gelifluction. Landforms resulting from this process are nivation hollows and nivation benches. Nivation hollows tend to be formed in sandy or silty rocks that are uniformly fine-grained. The resultant frost-shattered debris is therefore also fine-grained and is easily removed by meltwater. This leaves a hollow in which snow can accumulate to a greater depth. It is believed that some cirque glaciers have originated in nivation hollows. Where the rock debris produced by frost shattering is a mixture of fine-grained and coarse-grained material, the finer material can be carried away by meltwater but the coarse debris is left behind to form a bench or terrace (Fig. 1.48(*f*)). Nivation hollows tend to occur most often on slopes that are sheltered from the prevailing winds, where snow tends to accumulate.

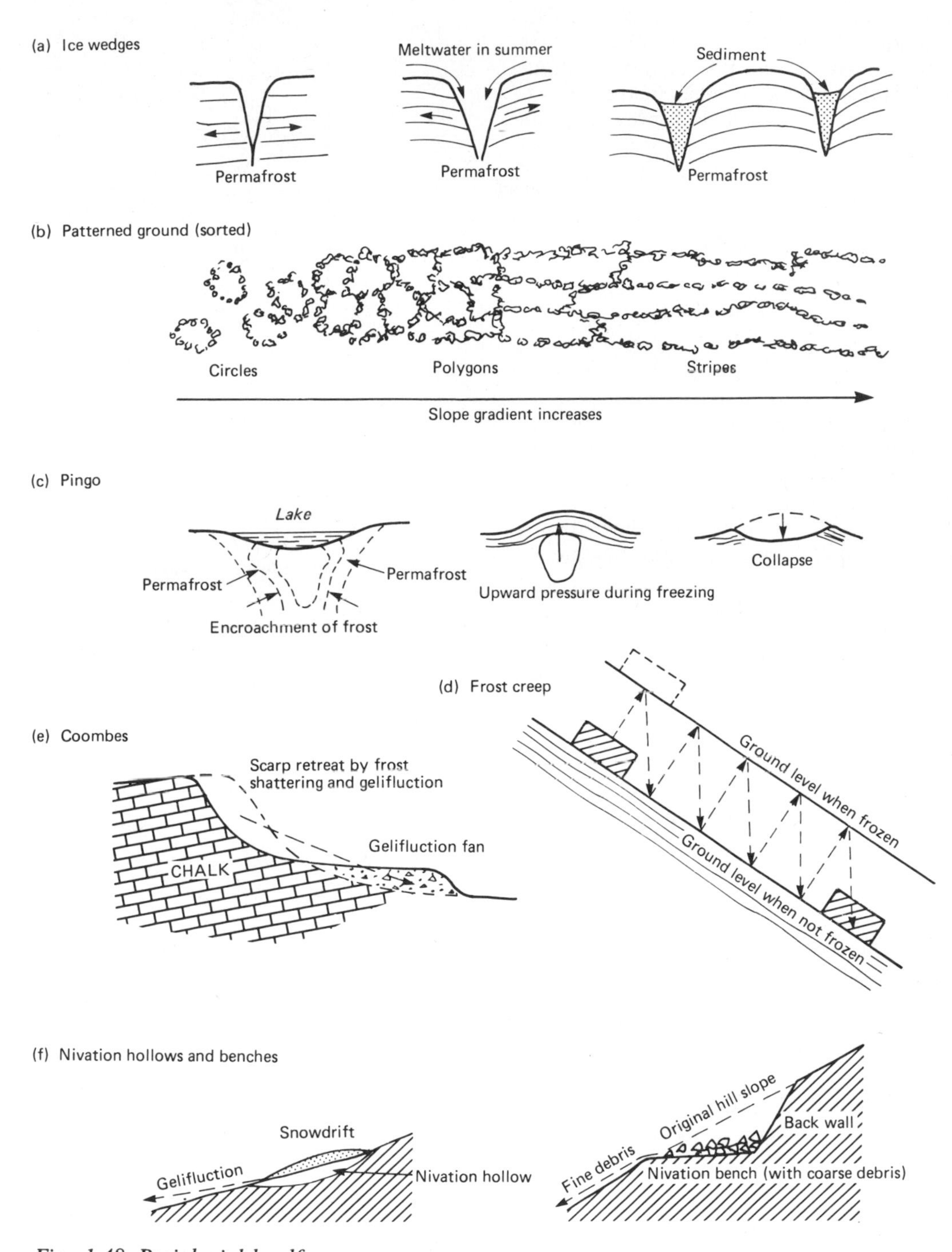

Fig. 1.48 Periglacial landforms

1.7 Coastal landforms

In general, the major landforms of coastlines have been created by past changes in the relative levels of the land and the sea. These have been responsible for the formation of, for example, offshore islands and major inlets of the sea. The detailed form of the coastline, however, is constantly being modified by the influence of waves and currents.

CHANGES OF SEA LEVEL

Many changes in the relative levels of land and sea have influenced the form of the world's coastlines during and since the Ice Age. They have occurred partly because the level of the sea has changed in relation to the land areas (*eustatic changes*) and partly because the levels of the land areas have changed in relation to sea level (*isostatic changes*). Every time a change has occurred waves and currents have recommenced their work of shaping the detailed landforms of the coastline, such as cliffs, wave-cut platforms, beaches, spits, etc.

Eustatic changes of sea level

Eustatic changes of sea level have been related to the expansion and contraction of the ice sheets during and since the Ice Age (page 65). In Pleistocene times there were four major periods when the ice sheets advanced. At these times the level of the sea was relatively low over the whole world. Rivers tended to adjust their long profiles to this low base-level by deepening their valleys (page 46). When the ice sheets retreated, sea level rose again and flooded the lower parts of these valleys. Hence, many of the world's rivers flow into estuaries which are usually termed rias if they occur in relatively hilly areas and are simply referred to as estuaries if they are in lowland areas. The drainage patterns of these river basins often tended to reflect the underlying geological structure (pages 32–34). Hence, the shape of the coastline also tends to be related to geological structure. Some coastlines trend generally parallel to the geological structures; others tend to cut across the structures (Fig. 1.49). Such flooded river valleys are common even in parts of the world that were never glaciated. In some cases, as sea level gradually rose, river estuaries became filled with sediment. Thus, no well-marked sea inlet was

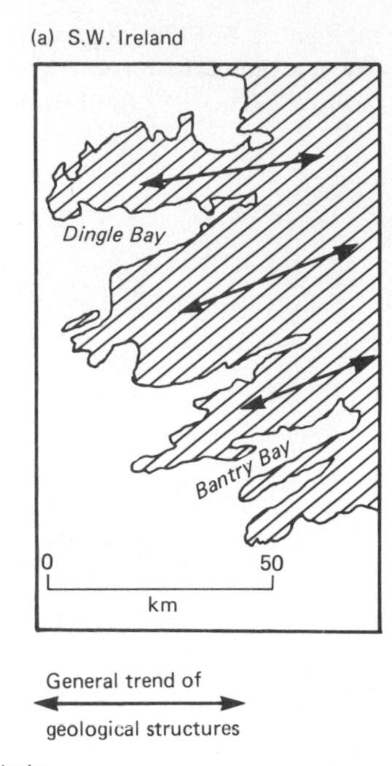

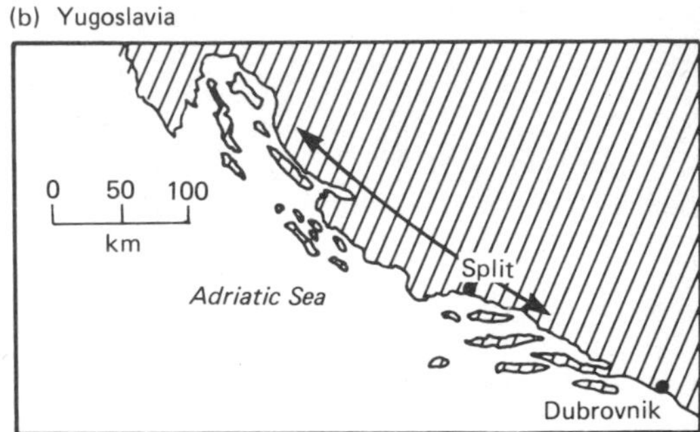

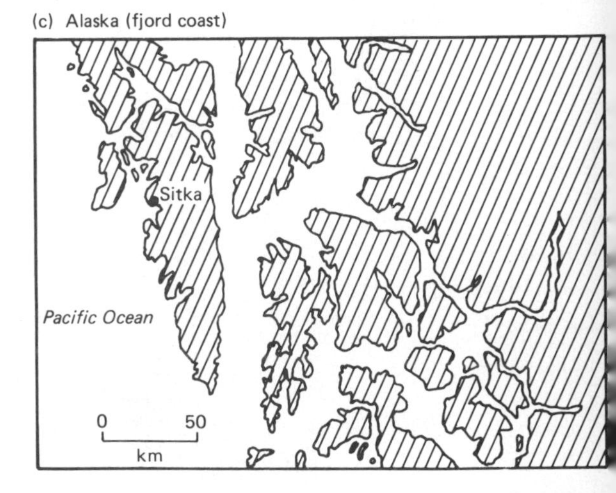

Fig. 1.49 Types of coastline

Howe Sound, a fjord in British Columbia

created. At the mouths of many English rivers such buried river channels exist. The estuaries have been filled with sediment to form a flood plain, across which the river meanders to the coast.

In parts of the world that were glaciated, fjords are commonly found on the coast. These are mainly the result of glacial erosion rather than a eustatic rise of sea level (page 69). Sea level can never have changed sufficiently to explain the depth of fjords.

Isostatic changes in the relative levels of land and sea

Areas that were covered by glaciers or ice sheets have had a more complex history. At the end of a glacial period the eustatic rise of sea level could be followed by a rise in the level of the land in relation to the sea that resulted from the removal of the weight of the ice (Fig. 1.50). In north-west Scotland there is an ancient sea beach (a 'raised beach') at a height of about 30 m above sea level (the '100-foot raised beach'). This raised beach is so high that it cannot be explained by reference to eustatic changes of sea level alone. This beach appears to have been created by marine erosion in a glacial period of the Ice Age. It only exists near the seaward ends of the sea-lochs, so it is thought that areas further inland were occupied by glaciers which prevented erosion by the sea. At the close of the Ice Age the eustatic rise of sea level has been followed by an isostatic rise of the level of the land as the weight of the ice has been removed. This theory is supported by the fact that present-day raised beaches tend to reach their greatest elevation in areas where it seems that accumulations of ice reached their greatest thickness. The '25-foot raised beach', for example, reaches its greatest height in north-west Scotland and gradually descends as it is traced southwards to sea level in the Irish Sea. This raised beach is thought to represent the maximum level of the sea in post-glacial times. Isostatic recovery has been greater in the north than in the south.

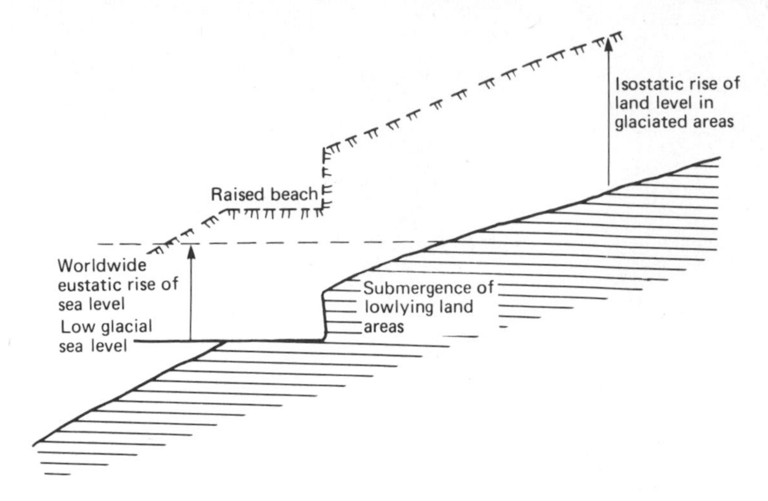

Fig. 1.50 Diagrammatic representation of changes in the relative levels of land and sea

COASTAL EROSION

Most of the erosional and depositional processes that occur on coastlines are the result of the action of waves.

Waves

Waves are approximately circular or elliptical movements of the water near the surface of the sea or any other body of water. They are caused by the friction exerted by winds as they pass over a water surface. When a wind passes over a water surface it tends to generate waves which move forward in the direction in which the wind is blowing, but there is little or no forward movement of the water itself. This tends to move slightly forward on the wave crests and backward in the troughs. Wave motion rapidly decreases with depth and there is little movement at a depth greater than half the wavelength. The height and the length of waves tend to increase if the wind speed increases. Very large waves can be generated if a strong wind blows for a long period of time across a very large water surface. The maximum distance of sea over which it is possible for a wind to blow towards a particular point on a coastline is known as the maximum fetch. If a strong wind blows from the direction of maximum fetch a stretch of coastline can receive extremely large waves. The direction of maximum fetch varies around the coast of Britain (Fig. 1.51(*a*)) from north or north-east on the east coast to south-west in the English Channel. The largest waves that reach a stretch of coastline, along the line of maximum fetch, are known as the dominant waves.

If the sea floor has a steep offshore gradient, vigorous waves may be able to reach the coastline with little loss of energy. A shallow sea floor, however, interferes with the forward movement of the wave and causes the wave-front to become curved (refracted). This may have the effect of concentrating the energy of the wave on particular parts of the coast. Shallowing of the sea bed often occurs to the seaward of headlands. In this case, refraction of the wave-fronts results in a concentration of wave energy on the headlands, thus tending to erode them back and to straighten the coastline (Fig. 1.51(*b*)). By the same process, waves approaching the coast obliquely are slowed as they enter shallow water and tend to become parallel to the coastline (Fig. 1.51(*c*)).

Erosion of the coast

An idealized representation of the process of coastal erosion is illustrated in Figure 1.51(*d*).

The waves first excavate a notch in the relatively steep coastal slope at about high tide level. Eventually, the rock above the notch collapses and a cliff and a wave-cut bench are created. Erosion continues by the pressure of the waves, the compression of air into crevices in the cliff, and the impact of eroded rock debris on the base of the cliff. Thus, the cliff retreats landwards and the wave-cut bench gradually becomes wider. Eroded material from the cliff is carried across the wave-cut bench by the backwash of the waves. Thus, the

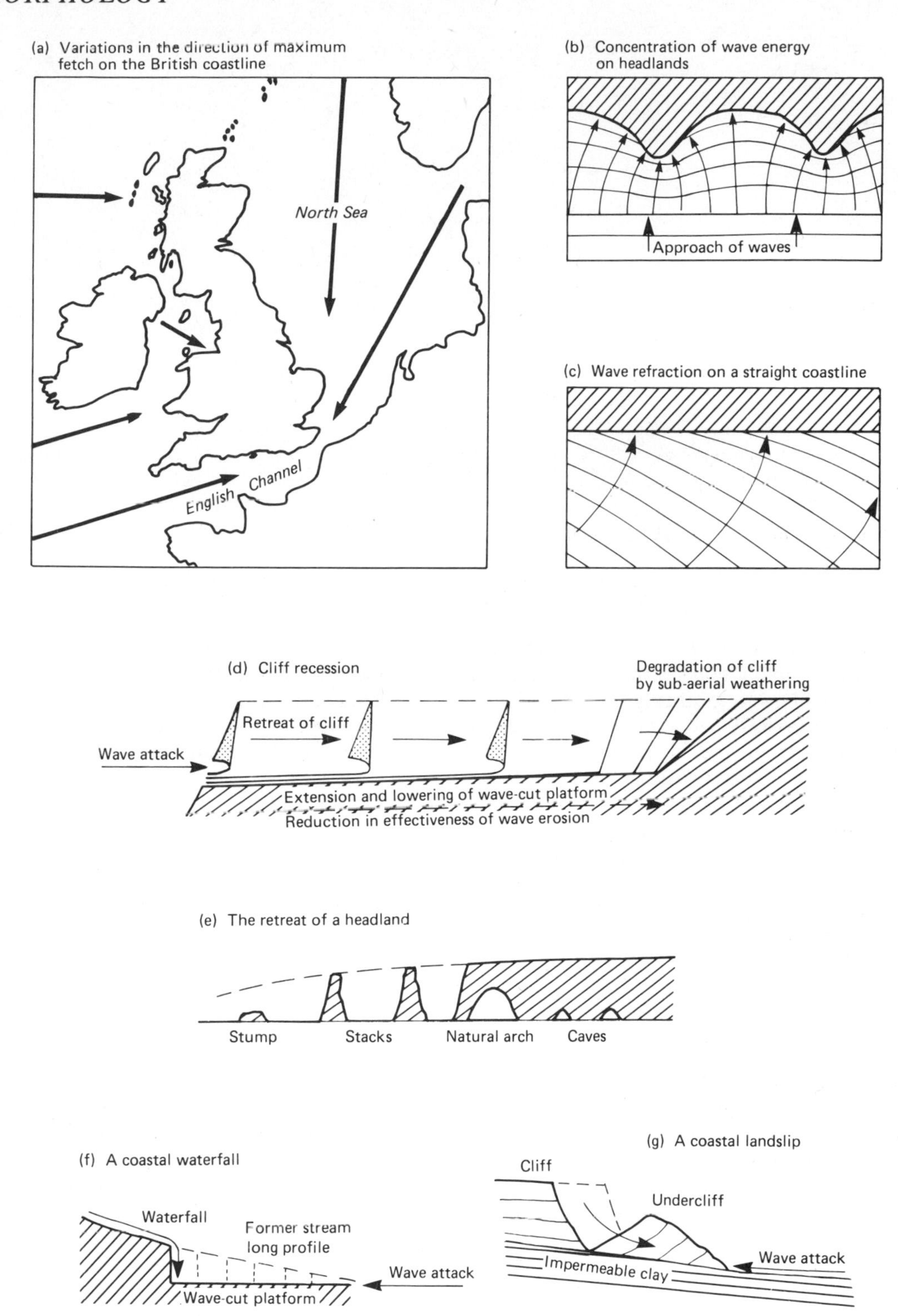

Fig. 1.51 Waves and wave erosion

A headland with stacks in the Isle of Wight

level of the wave-cut bench is gradually lowered. Most erosion usually takes place at high tide. As the wave-cut bench gradually widens, the effectiveness of wave erosion at the base of the cliff is reduced because of the increased friction that affects the waves in their passage across the wave-cut bench. Eventually the wave-cut bench becomes so wide and the vigour of the wave attack upon the cliff is so reduced that the cliff begins to be influenced mainly by sub-aerial weathering and its slope gradually declines. It becomes a 'dead' cliff with either a convex or a rectilinear profile, which soon becomes covered with vegetation.

The shape of a coastline, as represented on a map, can also be changed by processes of erosion. Sometimes a tendency exists for a coastline with bays and headlands to be straightened (Fig. 1.51(*b*)) as waves are refracted. Evidence of the erosion of a headland can be seen in features such as caves and natural arches, together with isolated stacks and stumps (collapsed stacks), indicating the headland's former extent (Fig. 1.51(*e*)). Adjacent to such headlands there may be bays in which deposition is taking place.

In other cases, however, erosion can cause the coast to become more indented. Waves seek out the zones of weakness and hollow out weak rocks to form coves. Erosion takes place along joints, faults and igneous intrusions to create caves and deep, narrow clefts known as geos. Great differences exist in the rate of retreat of the coastline in the face of attack by waves because of differences in the resistance offered by rocks. Cliffs of glacial drift in Holderness, for example, retreat particularly rapidly (up to 2 m per year) compared with the granite cliffs of Land's End.

In some cases, high cliffs have been observed to retreat more slowly in the face of wave attack than low cliffs. High cliffs will produce a greater amount of eroded debris that has to be removed by the waves' backwash before erosion can recommence. Thus, it seems likely that valleys reaching the coast will tend to be eroded to form bays, and the intervening ridges will form headlands. In some cases, erosion may be so rapid where a valley reaches the coast that a cliff is formed at its seaward end, so that it becomes a hanging valley from which a waterfall may descend to the beach. In this case the stream has been unable to deepen its valley sufficiently fast to match the rate of retreat of the coast (Fig. 1.51(*f*)).

The profiles of cliffs and wave-cut benches tend to reflect geological structure. Horizontally bedded chalk tends to produce vertical cliffs with horizontal wave-cut benches, as at Beachy Head. Strongly jointed granite, as at Land's End, tends to produce irregular castellated cliffs with no recognizable wave-cut bench at all, the whole coast having innumerable coves and stacks. Generally, if the coastal rocks dip inland or vertically, the cliffs tend to be nearly vertical and the wave-cut bench tends to have steep stack formations rising from it. If the strata dip towards the sea the cliffs tend to have a gentler slope. In this case, particularly where a permeable rock rests upon impermeable clay, coastal landslips can occur. Large masses of the permeable rock can slide, and sometimes rotate, over the wetted surface of the underlying clay (Fig. 1.51(*g*)). Good examples of landslip coastlines exist at Folkestone and in the south of the Isle of Wight.

COASTAL DEPOSITION

Deposition of sediment on coastlines takes several different forms. The most widespread is deposition by waves to create beaches, spits and bars. Deposition can also occur in sheltered locations, where there are few waves, to create salt marshes. Also, in some localities deposition by the wind has created large areas of sand dunes.

Deposition by waves

Beaches are accumulations of sand and shingle that often occur in sheltered positions on a coastline that is undergoing erosion, but they are also found in more exposed locations, covering a wave-cut platform, where a plentiful supply of sediment is available from the erosion of cliffs. An example of the former type of location is the development of beaches in the bays between headlands, where wave activity is less intense. The outer margin of a beach usually has a trend approximately at right-angles to the direction of approach of the dominant waves (Fig. 1.52(*a*)). The upper (inland) part of a beach tends to be steeper than the lower part, but the gradient varies according to recent wave conditions. During a period of strong onshore winds, large waves move inshore and, when they break, their force is directed steeply down on to the beach. Their backwash is very powerful, but their swash is weak. Destructive waves such as these tend to reduce the beach's gradient by transporting material seawards (Fig. 1.52(*b*)). Under quieter conditions, constructive waves, with a more elliptical orbit, tend to push material up the beach, thus steepening its gradient (Fig. 1.52(*b*)). These constructive waves can occur at various states of the tide, so they tend to build up beach ridges (*berms*) at various positions on the beach. In general, the upper part of a beach often has a fairly steep gradient and is often composed of coarse material such as pebbles. There may be a shingle ridge (a storm beach) above the level of the highest tides. Near the level of high tide there may be sets of beach cusps, particularly in locations where the waves tend to reach the shore at right-angles. The *cusps* are small, pointed 'headlands' usually a few metres apart (Fig. 1.52(*c*)). They are separated by smooth, rounded hollows, up to about a metre deep. The cusps are usually composed of shingle, but the hollows usually have a sandy floor. When the tide is high, the swash of the waves runs along the flanks of the cusps and the backwash tends to be concentrated in the hollows.

Shingle and sand can also be moved laterally along a beach by waves whose fronts approach the beach obliquely. This process is referred to as *longshore drifting*. The swash of the incoming waves carries shingle and sand up the beach obliquely, but the backwash carries them directly down the beach's steepest gradient (Fig. 1.52(*d*)). The direction of longshore drifting depends on the direction of approach of the dominant waves. On the east coast of England most longshore drifting is towards the south except near the Wash where it is westward. At a river mouth or at a bend in the coastline a spit can be built which in some cases diverts a river mouth. Orford Ness in Suffolk has diverted the mouth of the River Alde several kilometres to the south. Occasionally two spits may grow towards each other from opposite sides of an estuary. Many spits recurve at their distal ends. This can be caused by wave refraction (page 80) or possibly by interference from locally generated waves (Fig. 1.52(*e*)). As time passes, a spit tends to rotate so as to align itself at right-angles to the direction of approach of the dominant waves. This often involves the building of new laterals (Fig. 1.52(*f*)) and the erosion of older parts of the spit. This erosion may result in the exposure on the beach of the salt marsh that has developed in the shelter of the spit (Fig. 1.52(*f*)).

Other depositional landforms created by largely similar processes to those involved in the development of spits include *cuspate forelands*, like Dungeness, which have been influenced by two sets of 'dominant' waves coming from different directions, and *tombolos*, which are spits that have developed in such a way as to link an island with the mainland. Chesil Beach is a tombolo linking the Isle of Portland with the mainland. Wave action can also create *offshore bars*. These are landforms similar to spits but they are often situated some considerable distance offshore. They are believed to be formed by the erosion of the sea bed by waves that break some distance offshore. Hence, they can only exist in shallow seas. As the bar is driven inshore by the waves (Fig. 1.52 (*g*)) it emerges above the surface of the sea and may be colonized by plants which are able to trap wind-blown sand, so that the bar becomes a long, narrow island. It may develop laterals and when it has moved inshore it may be almost indistinguishable from a spit. The north coast of Norfolk has ideal conditions for the development of offshore bars – a shallow sea floor composed of soft, glacial deposits.

Other types of coastal deposition

Salt marsh development has been one of the most effective depositional processes on coastlines. It has created large new land areas in sheltered locations such as bays, river estuaries and on the landward side of spits and offshore bars. In such

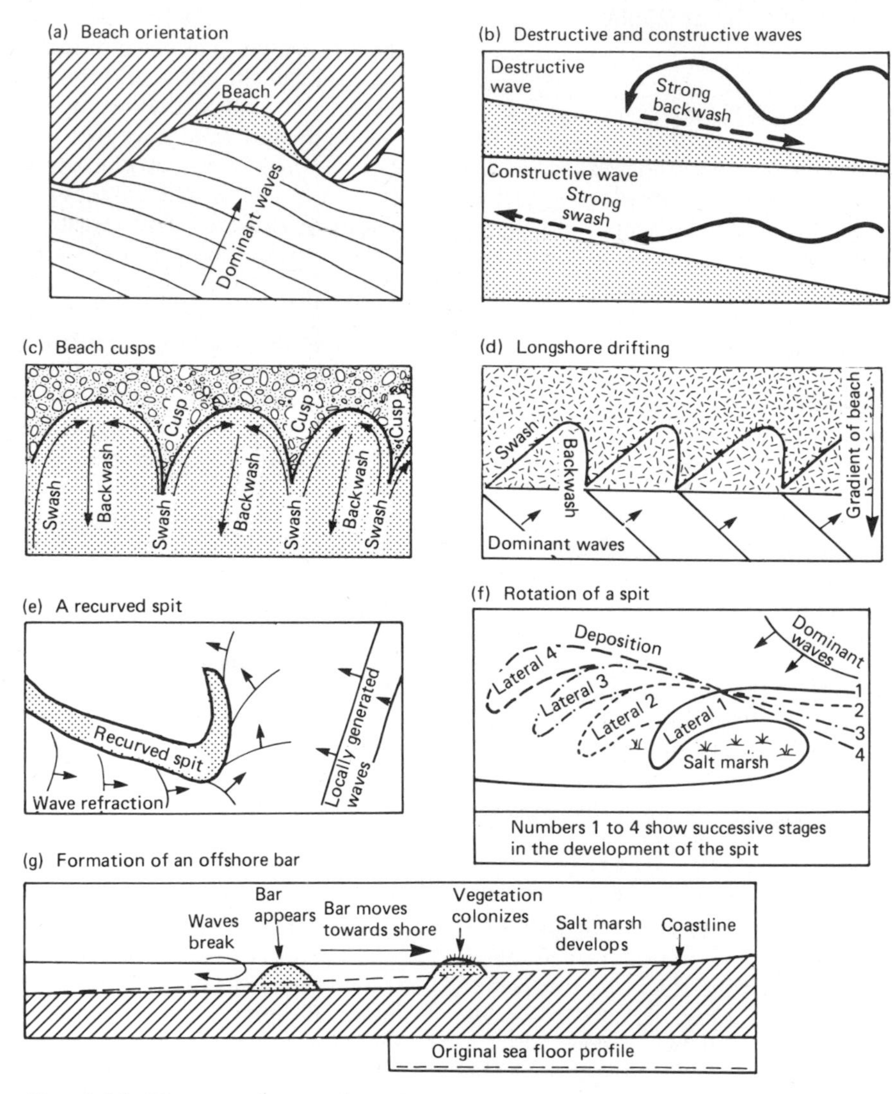

Fig. 1.52 Waves and wave deposition

Beach cusps on the west coast of Rhodes

sheltered areas the sea deposits sediment at high tide. As the tide ebbs the water drains away along randomly distributed channels. This produces a pattern of mostly linear depressions separated by low rises. Colonization of the higher areas by salt-tolerant vegetation then begins, notably by Salicornia and Suaeda maritima on the east coast of England. This vegetation increases the rate of sedimentation and other types of plant are able to colonize. Spartina townsendii first appeared in Southampton Water in the late nineteenth century and proved to be a most efficient trapper of sediment. It has encouraged salt marsh development along the south coast. Thus, the vegetated areas grow outwards, encroaching on the water channels until they are reduced to narrow creeks along which sea water moves according to the state of the tide, and some lateral erosion of the banks of the channels may occur (Fig. 1.53) causing slight subsidence. The marsh also grows upwards by further sedimentation until it is only rarely covered at high tide and may be reclaimed by man.

In areas where a wide, sandy beach is uncovered at low tide, in Britain especially on west or north facing coastlines, sand dunes commonly exist. At low tide, beach sand can be dried out sufficiently for strong winds from the sea to move large quantities towards the land, where it may be trapped by vegetation or areas of shingle. The accumulation of large dunes has been aided considerably by the growth of marram grass which traps wind-blown sand and has deeply penetrating root systems that can stabilize a dune. If there is a good supply of sand several lines of dune ridges can be created parallel to the coastline. In this case the dunes located furthest inland, the earliest to be formed, tend to decay, as less and less sand is able to reach them. If for some reason the vegetation on the dune decays strong winds may begin to erode the face of the dune, producing transverse channels known as *blow-outs*.

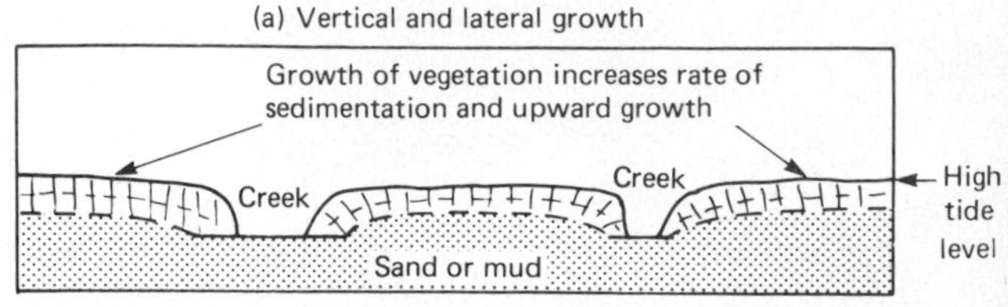

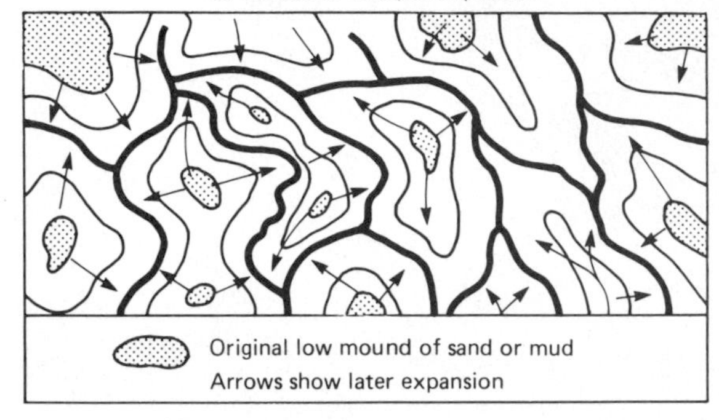

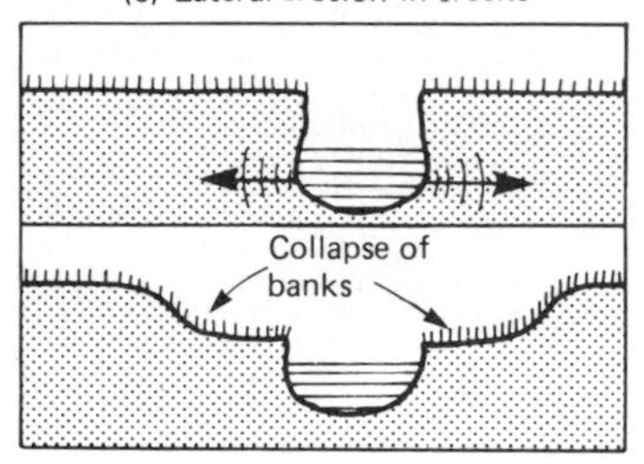

Fig. 1.53 Development of a salt marsh

Depositional layers in the salt marsh at Silverdale (Morecambe Bay)

Exercises

1. Comment on the following statements concerning the earth's major relief features.

(*a*) Many deep-focus earthquakes occur near island arcs and deep-sea trenches and many shallow-focus earthquakes occur near mid-oceanic ridges.

(*b*) In the south Atlantic the Mid-Atlantic Ridge is situated in longitude 10–15 degrees west. Volcanic activity still occurs at Ascension (longitude 15 degrees west). It used to occur, but has now ceased, at St. Helena (longitude 8 degrees west).

(*c*) Some evidence appears to exist that the Pacific plate is invading the Atlantic Ocean.

(*d*) In the Tibet Plateau and the Himalayas there are frequent earthquakes but there is no volcanic activity.

(*e*) Some oceanic volcanoes erupt mostly through a single vent.

(*f*) The mid-oceanic ridges have a much greater total length than the destructive plate margins.

2. Explain the extent to which the landforms created by weathering are a reflection of (*a*) rock type and (*b*) climate.

3. Discuss the factors that can influence the form of hillside slopes.

4. Referring to actual examples, explain the meaning of the terms 'feedback' and 'regulator' when

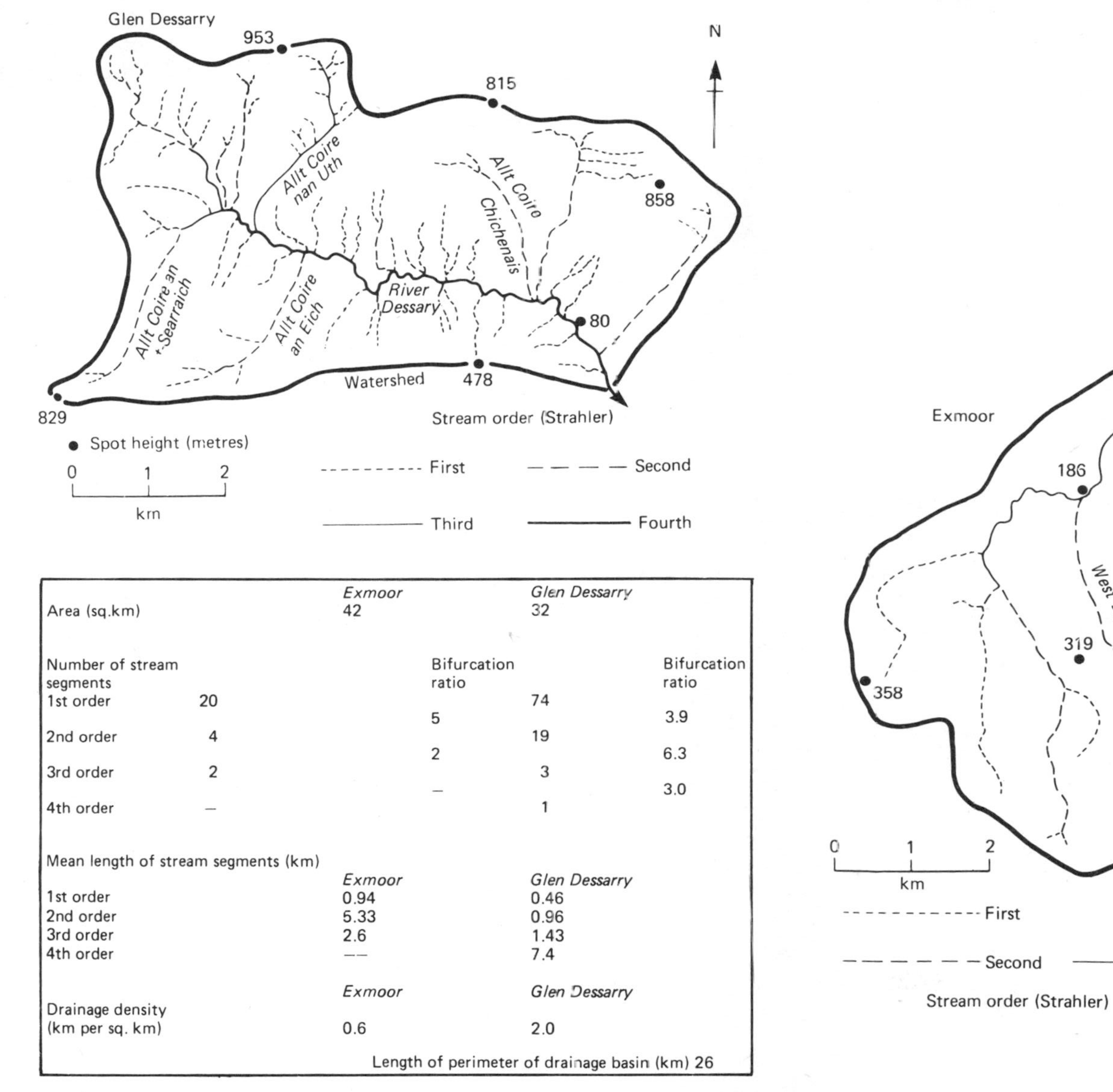

	Exmoor	Bifurcation ratio	Glen Dessarry	Bifurcation ratio
Area (sq.km)	42		32	
Number of stream segments				
1st order	20	5	74	3.9
2nd order	4	2	19	6.3
3rd order	2	—	3	3.0
4th order	—		1	

Mean length of stream segments (km)	Exmoor	Glen Dessarry
1st order	0.94	0.46
2nd order	5.33	0.96
3rd order	2.6	1.43
4th order	—	7.4

	Exmoor	Glen Dessarry
Drainage density (km per sq. km)	0.6	2.0

Length of perimeter of drainage basin (km) 26

Fig. 1.54 Characteristics of two drainage basins (for Question 6 on page 88)

they are used in relation to geomorphological systems.

5. Why are some drainage basins more likely than others to be affected by serious flooding?

6. Refer to the data provided for the Glen Dessarry drainage basin and a pair of drainage basins in Exmoor (Fig. 1.54).

(*a*) For the Glen Dessarry drainage basin calculate and comment on
 (i) its form factor;
 (ii) its compactness coefficient.

(*b*) Describe and attempt to explain the differences between the Glen Dessarry drainage basin and the two Exmoor drainage basins.

7. Why are so many river long profiles in the shape of a concave-upward curve? In what circumstances does this not occur?

8. Explain how volcanic activity can create (*a*) mountains; (*b*) plateaux; (*c*) surface depressions.

9. In what ways and for what reasons are the slope profiles of desert areas different from those in humid areas?

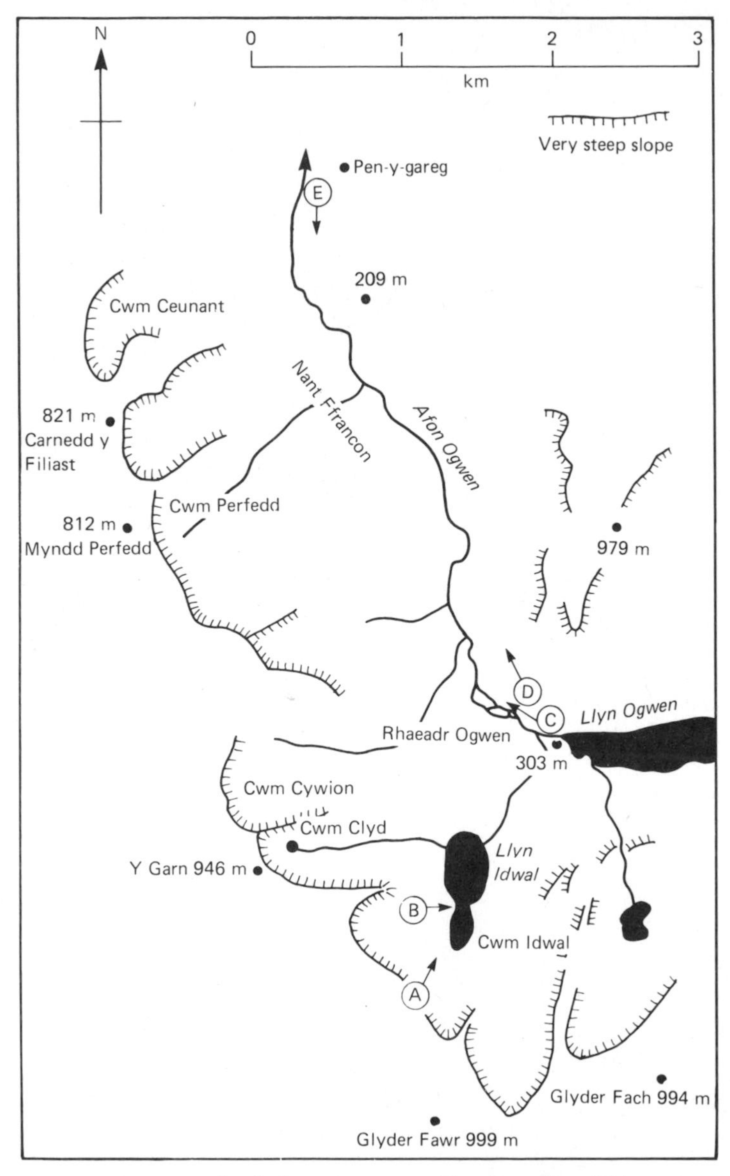

Fig. 1.55 Cwm Idwal: location map (for Question 10)

10. Photographs A, B, C, D and E below have been taken from the viewpoints labelled A, B, C, D and E respectively in Figure 1.55 (page 88). Using both the map and the photographs write an explanatory account of the landforms of this part of North Wales.

10A

10B

10C

10D

10E

11. (a) Describe the characteristics of the coastal landforms illustrated in photographs A to G on this page and page 91.

(b) In each case explain the processes that have been responsible for the creation of the landforms shown in the photograph.

11A

11B

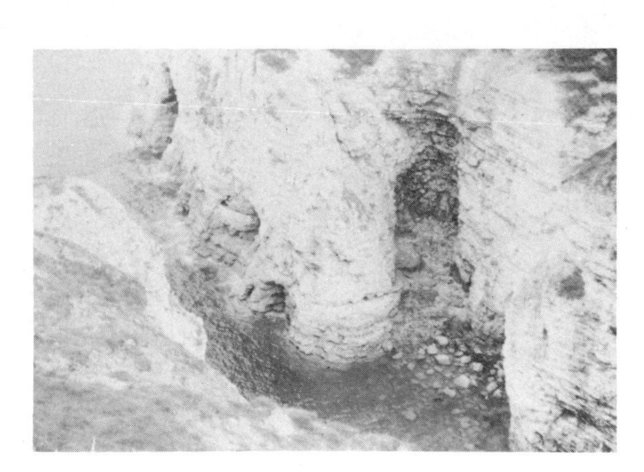

11C

11D

11E

11F

11G

2 Meteorology and Climatology

2.1 World temperatures

THE SOLAR ENERGY SYSTEM

The solar energy system comprises the sun, the earth and the earth's atmosphere. Energy travels from the sun in the form of electro-magnetic waves that are able to pass through empty space. The high temperature of the sun's surface allows it to generate visible light rays and also invisible infra-red and ultra-violet rays whose wavelengths are outside the range of visible light. Solar radiation can be transmitted through some substances such as air or water. It can also be reflected or absorbed. The absorption of solar radiation raises the temperature of the substance that absorbs it.

Figure 2.1 shows in a simplified form the characteristics of the solar energy system. The earth's surface and the atmosphere function as open systems and exchange energy with each other and with outer space. Such open systems tend to adjust themselves so that their output of energy is equal to the rate of input. Thus, the store of energy within each system tends to remain constant. This is known as dynamic equilibrium or a steady state. The attainment of a steady state by rivers is explained on page 43. In Figure 2.1, 100 units (27 + 24 + 45 + 4) of solar energy reach the atmosphere. Of these, 27 units are reflected or scattered back to space. This is chiefly the result of reflection from the upper surfaces of clouds. 24 units, mostly in the infra-red wavelengths, and mostly in the lower layers of the atmosphere, are absorbed by carbon dioxide and water vapour. This helps to raise the temperature of the lower atmosphere. 49 units of solar energy pass directly through the atmosphere and reach the earth's surface. Here, the surfaces of the land and the sea reflect about 4 units. These, together with the 27 units from the atmosphere, referred to above, make up the earth's *albedo*, the amount of solar energy that is reflected into space. Finally, 16 units are radiated directly back to space from the earth's surface. The largest transfers of solar energy take place between the earth's surface and the atmosphere. The earth's surface receives 97 units from the atmosphere, partly by the downward radiation of heat from clouds, and it transfers 126 units back to the atmosphere either as *sensible heat* (air being warmed by contact with the ground) or as *latent heat* (resulting from evaporation and condensation). The atmosphere then returns its surplus of 53 units to space. The tables of inputs and outputs for the atmosphere and the earth's surface (Fig 2.1) show that a condition of dynamic equilibrium exists, and average temperatures will tend to remain constant.

The earth's import of solar energy

Little loss of solar energy takes place as it travels across space to the outer edge of the earth's atmosphere. As it enters the atmosphere, however, it tends to suffer losses which begin first at the short-wave end of the spectrum. At an altitude of about 150 km ultra-violet radiation begins to be

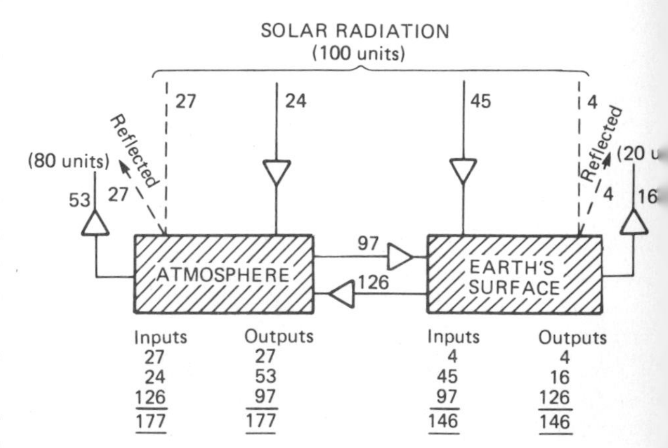

Fig. 2.1 The solar energy system

lost. Further down, visible light rays are deflected in various directions by molecules of gas, a process known as Rayleigh scattering. This is most pronounced at the short-wave (blue) end of the visible spectrum. The result is that a clear sky appears to be blue, because blue light reaches our eyes from all parts of the sky. Red light at the long wave end of the visible spectrum resists this scattering. Consequently the sun appears to be red at sunset because of the long, oblique traverse of sunlight through the atmosphere at this time. As solar radiation penetrates deeper into the atmosphere the infra-red part of the spectrum begins to be absorbed, especially by carbon dioxide and water vapour. The water vapour content of the air tends to vary from place to place and from time to time. Infra-red absorption is particularly effective when the sky is cloudy. In addition, the upper surfaces of clouds are able to reflect large amounts of solar radiation as part of the earth's albedo. A thick cumulonimbus cloud, for example, can result in hardly any solar energy reaching the ground surface. It can reflect up to 90% of all incoming solar energy.

The nature of the earth's surface also influences the proportion of solar radiation that is retained in the earth's surface and atmospheric subsystems. The surfaces of both land and water can reflect considerable proportions of incoming radiation. The albedo for a water surface is quite low if the sun is high in the sky but it increases greatly with a decrease in the sun's altitude, as at sunset for example, or in high latitudes. Snow-covered surfaces have a particularly high albedo. Forested areas tend to reflect comparatively little of the incoming solar radiation.

Interaction between the earth's surface and the atmosphere

The dominant feature of the solar energy system is the very high level of the exchanges of energy between the atmosphere and the earth's surface (Fig. 2.1). These result in striking variations in temperature in the *troposphere*, the lowest part of the atmosphere. The earth radiates energy at a relatively low temperature, as long-wave (heat) radiation. This infra-red radiation is, to a great extent, absorbed by water vapour, carbon dioxide, and clouds in the troposphere. Relatively little escapes directly to space. In turn, the troposphere re-radiates the terrestrial radiation that it has absorbed, partly to space and partly back to the earth's surface. This process has been likened to the heating of a greenhouse by the sun. The greenhouse glass, like the atmosphere, allows short-wave solar radiation to pass through, but it tends to hinder the escape of long-wave (heat) radiation.

Heat is also transferred from the earth's surface to the troposphere by convection. If the earth's surface becomes warm, the air that is in contact with it becomes heated, and rises to a higher level by convection. A vertical heat exchange can also occur as a result of water on the earth's surface being evaporated, causing the surface to lose heat. When condensation takes place, latent heat is released into the atmosphere. As a result of these processes, the temperature in the troposphere decreases with height at an average rate of 6.5°C per km. It should be stressed, however, that this is an average rate and it does not apply at all times. Under certain conditions it is possible for temperature to increase with height, a condition known as an inversion.

The troposphere is capped by such an inversion (Fig. 2.2), where temperature first remains constant with increasing height and then begins to increase. The inversion at the top of the troposphere is at a height of about 16 km at the equator and 8 km at the poles. This inversion marks the upper limit of weather variations in the atmosphere. It is known as the *tropopause*. Above the tropopause, in the *stratosphere*, temperature generally rises with increasing height (Fig. 2.2). In this zone ultra-violet radiation from the sun is ab-

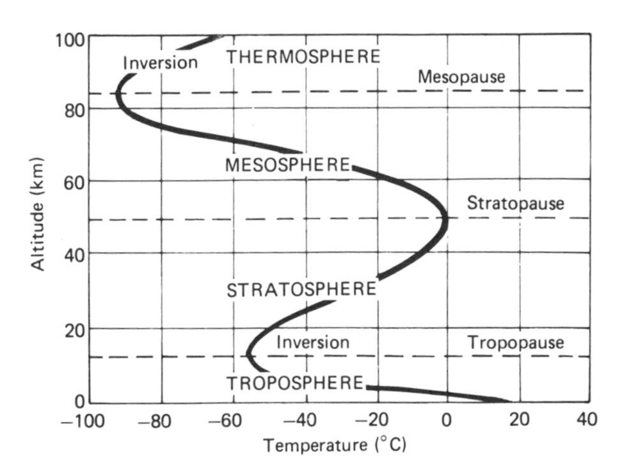

Fig. 2.2 Temperature variations with altitude in the atmosphere

sorbed by ozone. At the upper limit of the stratosphere, the *stratopause*, at about 50 km, the temperature may be just above freezing point. Above this level, temperatures decline with elevation in the *mesosphere* but rise again as height increases in the *thermosphere*. Within the troposphere it is possible for the lapse rate of temperature with altitude to vary for a number of different reasons. In anticyclones (centres of high pressure) the air is descending and becoming warmer through compression. This can result in a considerable depth of air being warmed, giving an inversion if the ground surface cools down at night under the relatively clear skies. Alternatively, the lower troposphere may be cooled by coming into contact with a snow-covered ground surface. In areas of varied relief cooled air can collect in valley floors in winter. In contrast to anticyclones, the air in low pressure systems tends to cool by expansion as it rises. This tends to increase the temperature gradient instead of producing an inversion.

TEMPERATURE VARIATIONS AT THE EARTH'S SURFACE

The influence of latitude

Latitude influences temperatures in two main ways. It affects the angle at which solar radiation passes through the atmosphere and also the angle at which this radiation strikes the earth's surface (Fig. 2.3). It also influences the relative lengths of day and night over the earth's surface (Fig. 2.4). The earth's surface can only absorb solar energy during daylight hours.

Figure 2.3 illustrates the variations in the angle of incidence of solar radiation at noon at different latitudes. The intensity of insolation is greatest when the sun's elevation is greatest. It is clear that the sun's elevation at noon is highest on the average at the equator. It is directly overhead twice a year, at each equinox (Fig. 2.3(*a*)), and, since the noonday overhead sun does not extend beyond the Tropics of Cancer and Capricorn (latitudes 23½°N and 23½°S) (Figs 2.3(*b*) and (*c*)), the minimum altitude of the noonday sun at the equator can never be less than 66½ degrees (90°–23½°). Such a high angle of incidence of solar radiation means that solar energy takes a very short route through the atmosphere so that relatively little is reflected to space and it is concentrated in relatively small areas of the earth's surface. At all points within the tropics the noonday sun is overhead twice a year but at the Tropic of Cancer it is overhead only at the June solstice (Fig. 2.3(*b*)) and at the Tropic of Capricorn only at the December solstice (Fig. 2.3(*c*)). One would therefore expect temperatures to be generally high within the tropics. At the Tropics of Cancer and Capricorn, however, one would expect the annual range of temperature to be greater than at the equator since the altitude of the noonday sun varies from 90 degrees at the summer solstice to only 43 degrees (90°–47°) at the winter solstice (Figs. 2.3(*b*) and (*c*)).

Outside the tropics, both northwards and southwards, the elevation of the midday sun steadily decreases in both summer and winter and it differs by 47 degrees between the June and December solstices. Eventually, at the Arctic Circle at the December solstice (Fig. 2.3(*c*)) and at the Antarctic Circle at the June solstice (Fig. 2.3(*b*)), the sun does not rise above the horizon and even at the summer solstice it only reaches an elevation of 47 degrees. At the North Pole and the South Pole the sun is below the horizon for half the year and above the horizon (continuous daylight) for the other half, but even at the summer solstice it only rises 23½ degrees above the horizon.

Figure 2.4 illustrates the influence of latitude upon the varying lengths of day and night. In each hemisphere the length of daylight increases polewards in the respective summer season until at the Arctic and Antarctic Circles, at the respective solstice, there is one day with continuous daylight. Similarly, in the winter season, the length of daylight decreases polewards until, at the Arctic and Antarctic Circles there is one day on which the sun never rises. The poles have continuous daylight for half of the year and continuous darkness for the other half, 'dawn' and 'sunset' occurring at the equinoxes.

Because of the factors described above a broad belt between latitudes 35°S and 40°N absorbs a greater amount of solar energy than it loses by radiation and the rest of the world radiates more energy than it absorbs. But the areas with a surplus in low latitudes do not become hotter and hotter, nor do the deficit areas become colder and colder. This is because heat is transferred from low to high latitudes by two processes. Water vapour is evaporated in low latitudes and transported polewards by warm air masses. When condensation occurs latent heat is released to become sensible heat. Heat from low latitudes is also transported polewards in warm ocean currents such as the Gulf

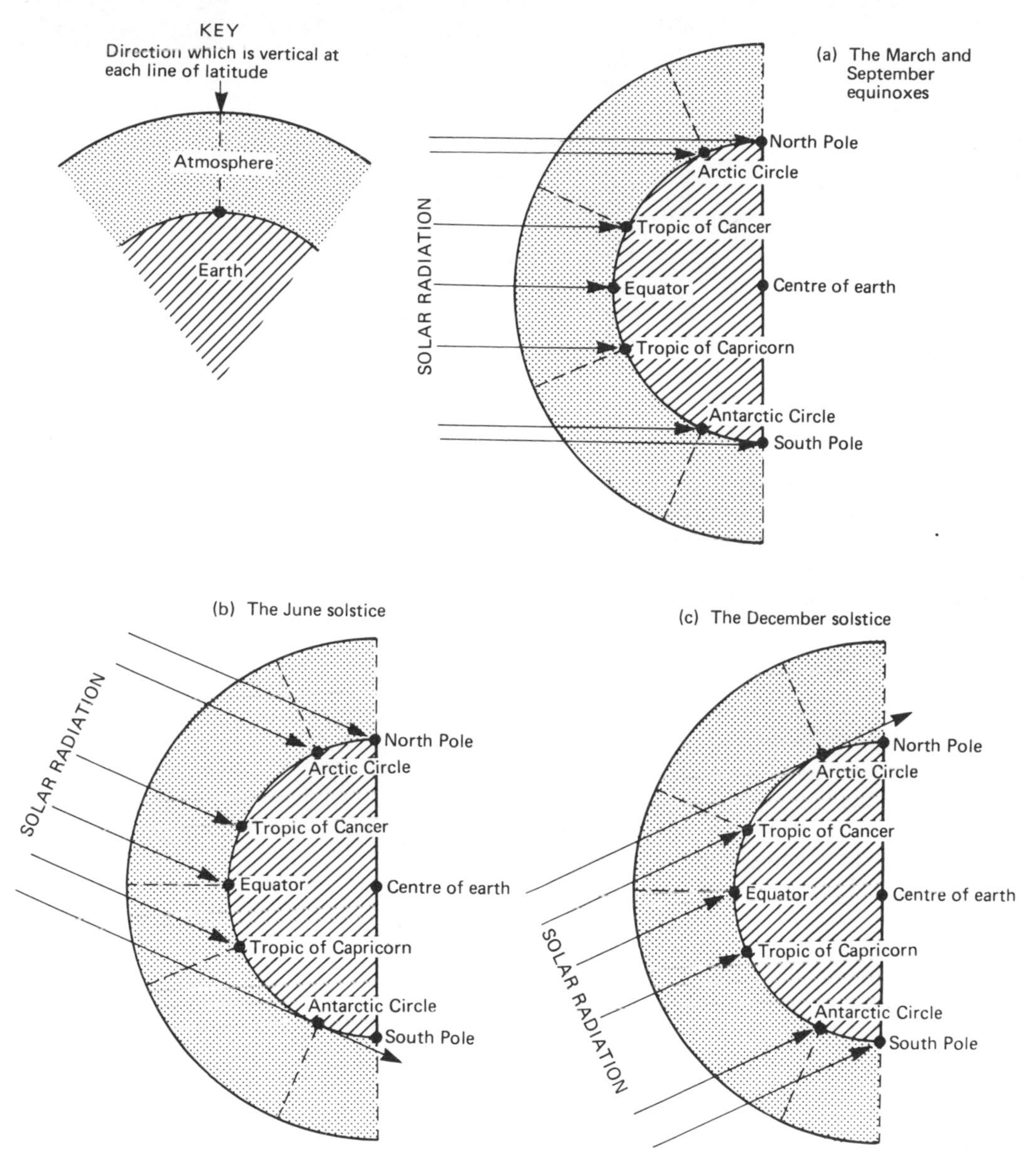

Fig. 2.3 Seasonal changes in the earth's attitude towards the sun (not to scale)

Stream and the North Atlantic Drift. If this transfer of heat did not occur equatorial areas would be over 10°C hotter and polar areas would be over 20°C colder.

In general, world temperatures decline steadily from the tropics towards the high latitudes (Fig. 2.5) and the annual range of temperature increases polewards from less than 10°C in the tropics to well over 40°C in continental areas of the northern hemisphere. This is partly the result of the poleward increase in the difference between the summer and winter rates of absorption of solar energy.

Despite the transfer of heat from low to high latitudes described above, the isotherms on a world temperature map (Fig. 2.5) trend broadly from east to west and demonstrate that solar radiation is an important cause of temperature differences over

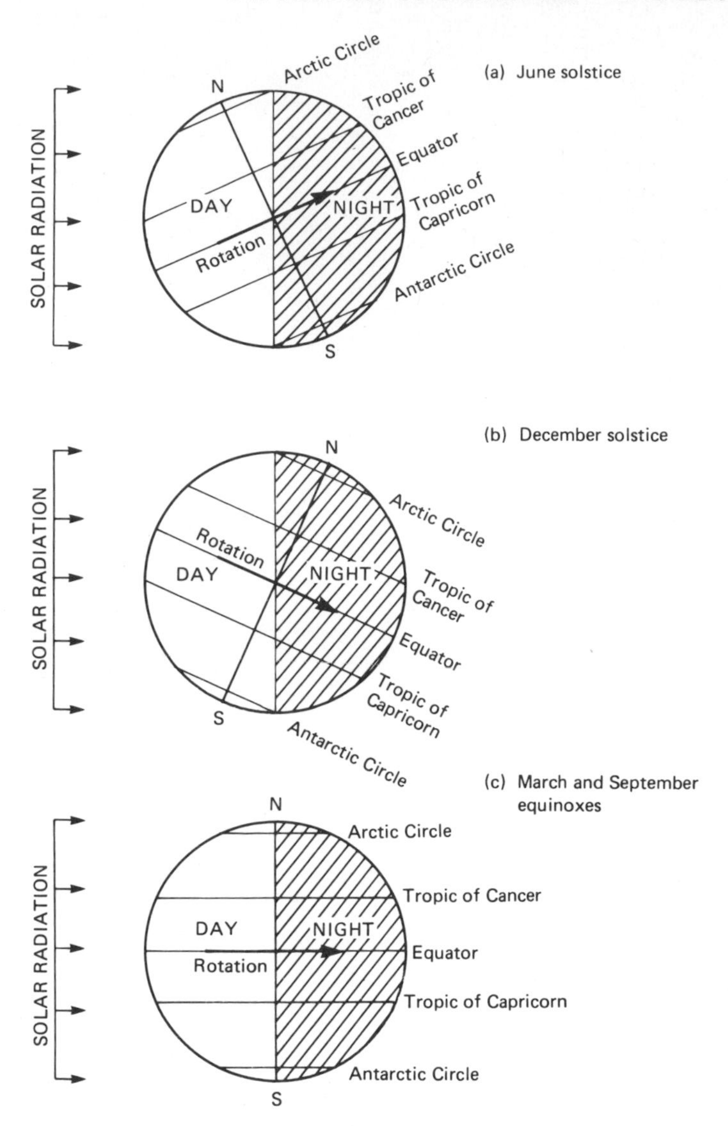

Fig. 2.4 Diagrammatic representation of varying lengths of daylight

the earth's surface. This east-west trend of the isotherms is most evident in the southern hemisphere to the south of the Tropic of Capricorn where the earth's surface is predominantly water (Fig. 2.5).

Continental and oceanic influences

Another major influence on the pattern of world temperatures is the distribution of land and sea over the earth's surface. For a number of reasons land and water surfaces differ greatly in their capacity to absorb and re-radiate solar energy and consequently in their ability to influence the atmosphere's temperature. When solar energy reaches a water surface it can penetrate to a considerable depth. Also, the water is constantly moving and mixing warmed water with cooler water. It is therefore unlikely that the water's temperature will rise at all quickly unless it is very shallow. When solar energy reaches a land surface, on the other hand, it is absorbed in a very shallow surface layer that can reach a very high tempera-

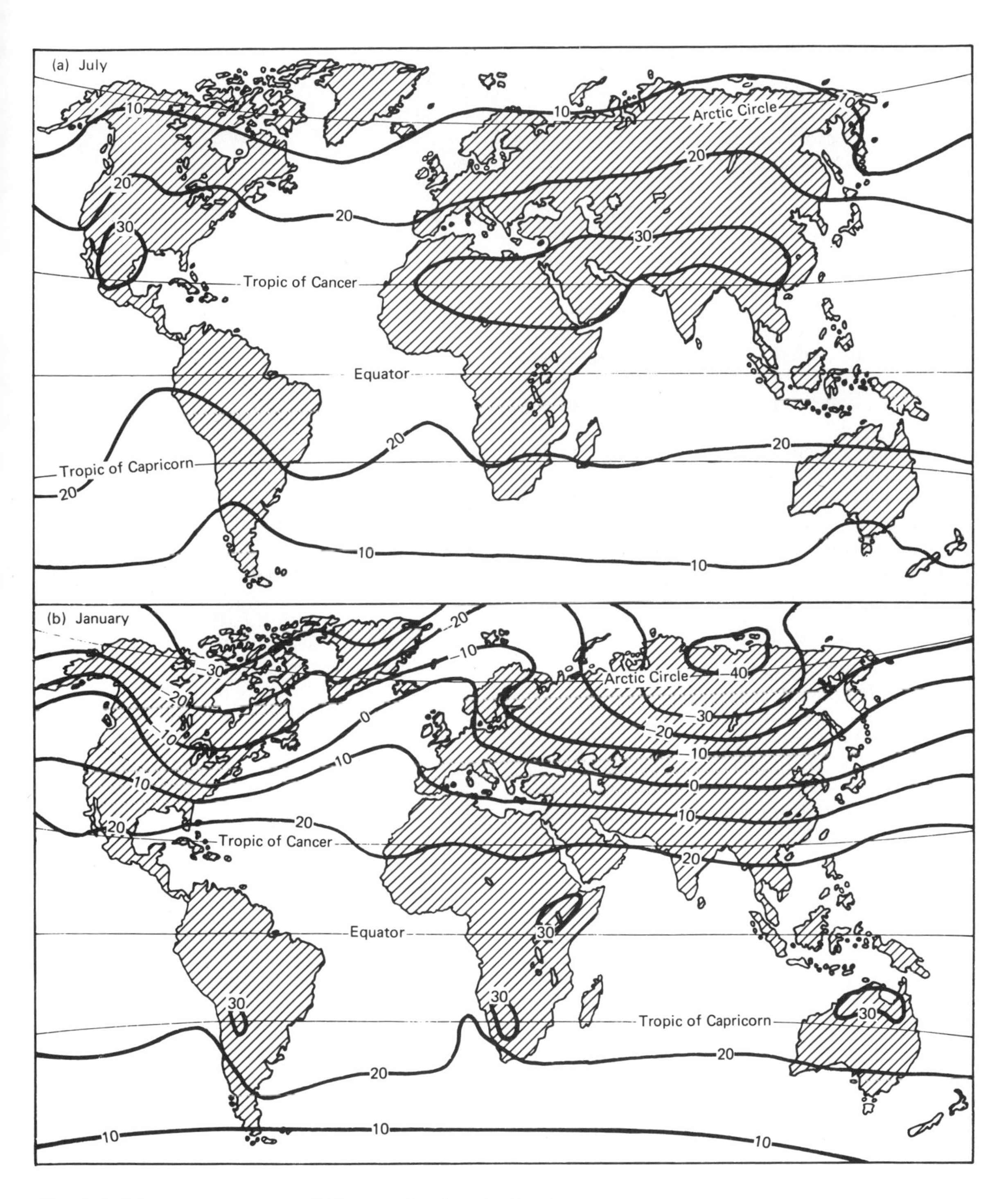

Fig. 2.5 Mean temperatures (°C) (sea-level equivalent)

ture. Also, evaporation can usually occur more readily over a water surface than over land. Evaporation has a cooling effect as latent heat is taken up. A much greater amount of heat is needed to raise the temperature of a water surface than a land surface. The specific heat of water is greater than that of land.

Thus, as solar radiation increases as the summer season approaches, temperatures in continental areas rise more quickly and eventually reach a higher level than in oceanic areas in the same latitude. When solar radiation is decreasing during autumn and winter, the temperature in continental areas declines more quickly and falls to a lower level than in oceanic areas. The continents therefore tend to have warmer summers and colder winters than the oceans. Also, in continental areas the warmest month of the year is usually July in the northern hemisphere and January in the southern hemisphere whereas, in oceanic areas, the warmest months are often August and February respectively.

Figure 2.5 illustrates the principles outlined above. In the southern hemisphere, where there is much more ocean than land, the isotherms in both July (winter) and January (summer) run generally parallel to the lines of latitude and it is clear that the relatively small land areas have little effect on temperatures apart from producing three isolated areas with mean January temperatures of over 30°C in January (summer). In the northern hemisphere, however, the continents of North America and Eurasia have a great effect on temperatures. In July (summer) (Fig. 2.5(*a*)) the isotherms tend to curve northwards over these continents and southwards over the intervening oceans, indicating that, in any latitude, the continents are warmer than the oceans. The two large areas with mean July temperatures of over 30°C hardly ever reach the sea coast. In January (Fig. 2.5(*b*)) the opposite occurs. The isotherms bend southwards over the continents and northwards over the oceans, showing that the oceans, particularly in higher latitudes, are very much warmer than the continents. This temperature difference is much greater in winter than in summer, the curves of the isotherms being much greater in winter. Along the Arctic Circle, for example, mean temperatures vary from 0°C in the north Atlantic, between Iceland and Norway, to almost −40°C in north-east Asia. This contrast is illustrated in Figure 2.6. Reykjavik and Verkhoyansk are both situated very near to the

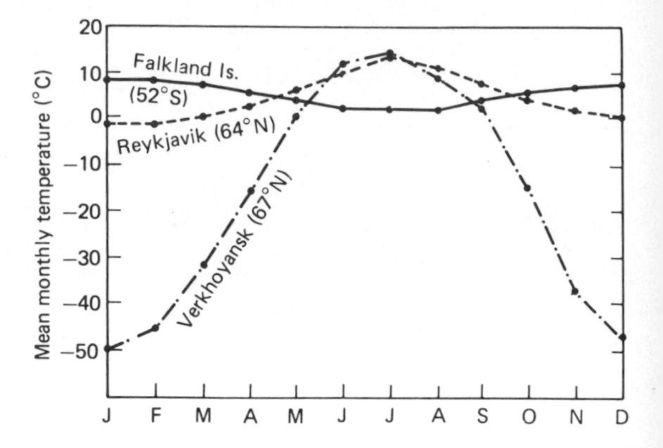

Fig. 2.6 Continental and oceanic temperature regimes

Arctic Circle, Reykjavik being located in Iceland in the north Atlantic Ocean and Verkhoyansk being in north-east Asia. Moving southwards in the northern hemisphere these temperature differences between ocean and continent tend to become smaller. The contrast is so great near the Arctic Circle partly because the earth has such a large land area near the latitude of the Arctic Circle and this is responsible for extremely low winter temperatures. In Figure 2.6 the Falkland Islands illustrate the very low annual range of temperature that is characteristic of southern hemisphere locations. Temperatures here are also generally rather lower than might be expected at a similar latitude in the northern hemisphere. Temperatures in the Falkland Islands are similar to those experienced in Iceland, yet Falkland is over 10 degrees nearer to the equator.

The influence of winds and ocean currents

The continental and oceanic influences on world temperatures explained above are generally accentuated by the influence of major pressure and wind systems and ocean currents.

The surplus of heat that tends to occur in tropical and subtropical areas (page 94) is transferred to higher latitudes that have a heat deficit. This is achieved by the movement of air (winds and air masses) and water (ocean currents). Warm maritime Tropical (mT) air originates mainly in the trade wind belt and the subtropical high pressure systems located at about latitude 30° north and south (Fig. 2.7(*a*)) and is carried to higher latitudes by the westerlies and their as-

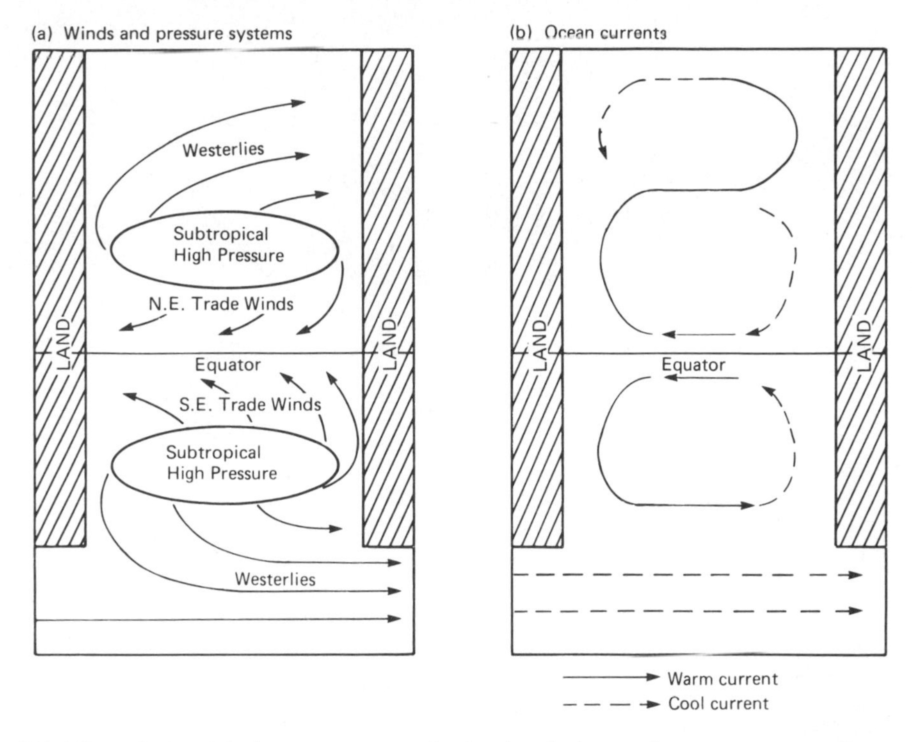

Fig. 2.7 The relationship between atmospheric circulation and ocean currents (generalized)

sociated depressions. In the northern winter, cold air masses consisting of continental Polar (cP) air can originate in the extremely cold northern continents of Asia and North America. These often invade the east coasts of the two continents and bring excessively low temperatures. The west coasts of these continents are more likely to experience invasions of maritime Polar (mP) and maritime Tropical (mT) air that has crossed an ocean and therefore brings much higher temperatures. The extent to which the maritime air masses are able to penetrate the continents depends on the location and the alignment of topographical barriers. In Europe, except for Scandinavia, there are few mountain ranges lying across the path of these maritime air masses. The Alps and the Pyrenees, for example, are aligned generally from east to west. In North America, however, the western cordilleras run along the whole length of the west coast and effectively block the entry of relatively warm air masses from the Pacific Ocean. In western Canada the mean January temperature falls from 3°C on the west coast at Vancouver to −15°C at Edmonton which is about 800 km inland. In Europe, on the other hand, the mean January temperature falls only from 5°C to 1°C between south-west Ireland and Bremen in West Germany, a distance of over 1000 km.

Ocean currents also tend to accentuate the temperature differences between coastal areas and continental interiors by reducing coastal temperatures in summer and increasing them in winter. Figure 2.7(*b*) shows an idealized pattern of ocean currents in the Atlantic and Pacific Oceans which both extend over a great latitudinal range. Their direction of flow is controlled partly by the prevailing winds, partly by the configuration of the coastline and partly by the earth's rotation (*Coriolis force*) which tends to deflect freely moving ocean currents to the right in the northern hemisphere and to the left in the southern. In both the Atlantic Ocean and the Pacific Ocean the north-east and south-east trade winds tend to create a westward flow of water along the equator (North and South Equatorial Currents) often with a reverse flow (Equatorial Counter Current) between the two. At the western side of the ocean the North and South Equatorial Currents are de-

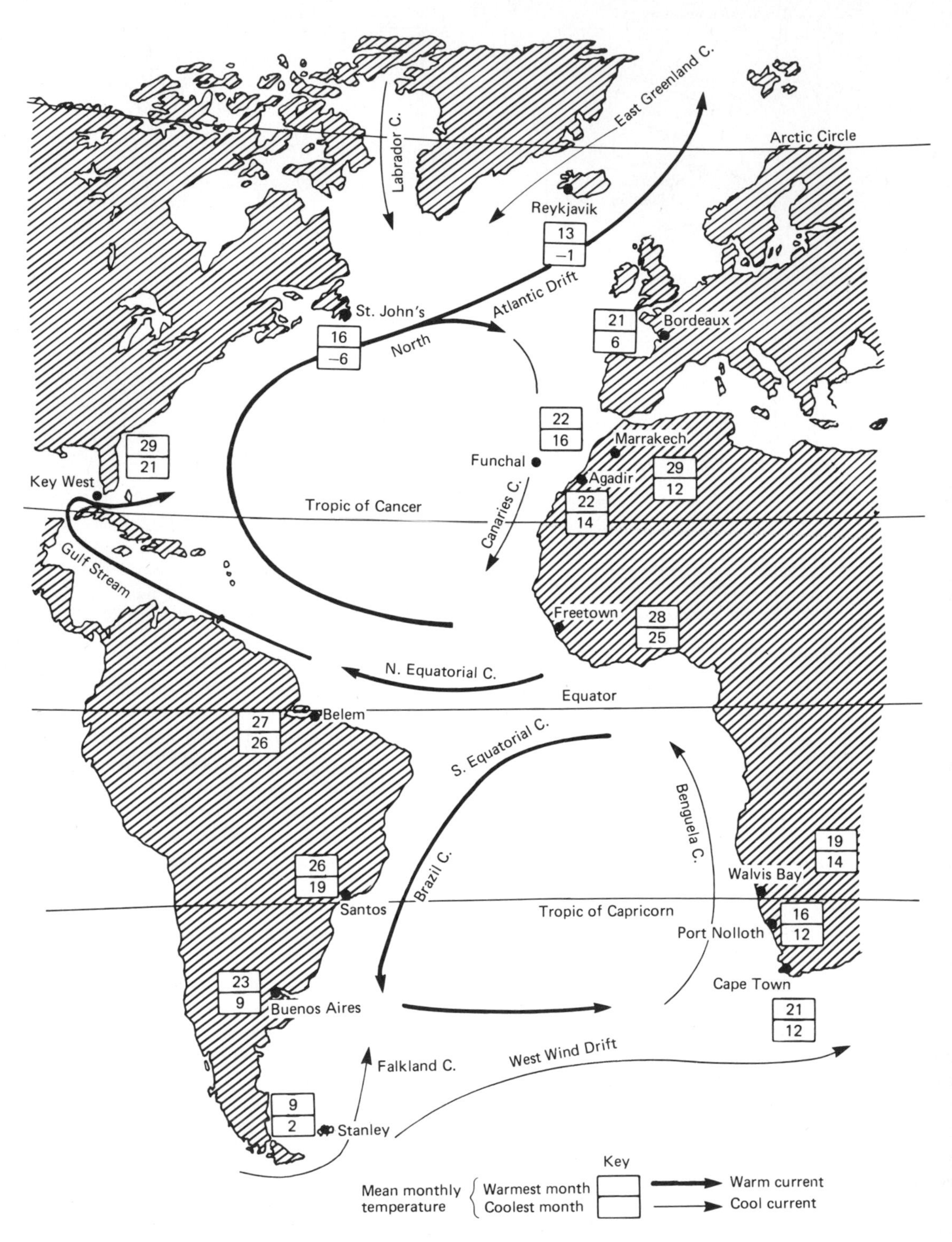

Fig. 2.8 The effect of ocean currents on the temperatures of the coastlands of the Atlantic Ocean

flected to the north and south respectively. This occurs partly because of the configuration of the coastline, particularly in the Atlantic Ocean (Fig. 2.8) and partly because of the winds circulating around the subtropical high pressure systems (Fig. 2.7(*a*)). The ocean currents then flow along the poleward sides of the subtropical high pressure systems in an easterly direction. When they reach the eastern coastline of the ocean they divide again. One branch returns along the coast to the equator, thus completing the circulation around the subtropical high pressure systems. As this current flows equatorwards it commonly causes an upwelling of cold water from the ocean depths. In the northern hemisphere the other branch runs along the eastern margin of the ocean (the west coast of the continent to the east) and, according to the configuration of the coastline, returns southwards at the opposite side of the ocean. In the high latitudes of the southern hemisphere, unlike the northern hemisphere, there is very little land, so, on the poleward side of the subtropical high pressure systems, a great drift of water moves from west to east around the earth's circumference, with little interference by land areas (Fig. 2.7(*b*)). Details of the oceanic circulation in the Atlantic Ocean are shown in Figure 2.8.

The various warm and cool ocean currents of the Atlantic Ocean have a striking influence on the temperatures of the coastal areas (Fig. 2.8). In the north, Bordeaux is much warmer than St John's at both seasons, but particularly in winter. St John's appears to receive little warmth from the North Atlantic Drift. Its coolest month is colder even than that of Reykjavik which lies much further north. In the far south, Stanley has relatively low temperatures at both seasons but its winters are not so cold as those of St John's, there being no large continent near Stanley. Near the equator temperatures on both sides of the Atlantic (at Belem and Freetown) are very similar. The chief contrasts in temperatures occur in the tropics and subtropics where there is a warm ocean current on the west side of the Atlantic and a cool current on the east side. Here, Agadir is 7°C cooler than Key West at both seasons and Walvis Bay and Port Nolloth are considerably cooler at both seasons than Santos. This cooling effect seems to be particularly well marked in summer, since, at that season, both Walvis Bay and Port Nolloth are cooler than Cape Town, despite being nearer the equator. An interesting illustration occurs in north-west Africa of the effect of the cool Canaries Current. The temperature range gradually increases from Funchal in Madeira, about 650 km from the coast, to Agadir, on the coast and then to Marrakech, which lies about 150 km inland. Even though Marrakech is fairly close to the sea it receives little of the cooling effect of the Canaries Current. Its warmest month is comparable with that of Key West, but it is considerably cooler in winter than Funchal.

Temperature variations at the local scale

The factors discussed above (latitude, the distribution of land and sea, winds and ocean currents) produce broad temperature variations over a large area, such as a country or a continent. Very considerable temperature variations can also occur at the local scale, over distances of a few kilometres or even less.

Relief features are an important cause of local temperature variations. In general, as the altitude of the land increases, temperatures near the ground tend to fall. This is because, as the atmosphere's density decreases with greater height, less of the earth's outgoing radiation can be absorbed by the air. At high altitudes, if the land surface is covered with snow which has a very high albedo, much of the incoming solar radiation is reflected. Daytime temperatures can rise to very high levels in deep, narrow valleys where solar radiation can be reflected from the valley sides into the valley floor and the heated air accumulates because its movement is restricted. At night, however, when the sky is clear, and particularly if the ground is covered with snow, the ground surface is rapidly cooled by radiation and consequently the air that is in contact with the ground is also cooled. This cold air flows down hill slopes to accumulate in the valley floors. Under these conditions night temperatures in the valley floors can be lower than those on the hill summits, thus forming an inversion (page 94).

Another important consideration relating to the temperature of hill slopes is the direction in which they slope (their aspect). The aspect of a slope influences the intensity of the solar radiation that it receives and also the length of time that it can be exposed to such radiation. In areas of strong relief considerable temperature variations may occur, particularly in east-west trending valleys. These are related to the position of the sun in the sky at the various seasons of the year. Figure 2.9 refers to such a valley in the mid-latitudes of the northern hemisphere. At location P (Fig. 2.9(*a*)) at the summer solstice in June the sun rises considerably to the north of east (R1) and sets considerably to the north of west (S1), and there may be well over 12 hours of sunshine in a day. At the equinoxes the sun rises in the east and sets in the west. At the winter solstice the possible duration of sunshine is much shorter and the sun rises well to the south of east and sets well to the south of west. Figure 2.9(*b*) shows the variations in the noonday sun's elevation at different times of the year. Its highest elevation is at the June solstice. At this time insolation on a level surface is more intense than at any other time. Figure 2.9(*c*) illustrates the application of these principles to a steep-sided, east-west trending valley. At noon the southern side of the valley casts a shadow across the valley floor which varies in width according to the season. It is narrowest at the June solstice but it widens to extend completely across the valley floor at the December solstice, only the higher part of the northern valley side being in direct sunlight. Thus, one would expect the northern valley side (the adret slope) to have generally higher temperatures than the southern (the ubac slope). In such conditions the adret slope is often preferred for

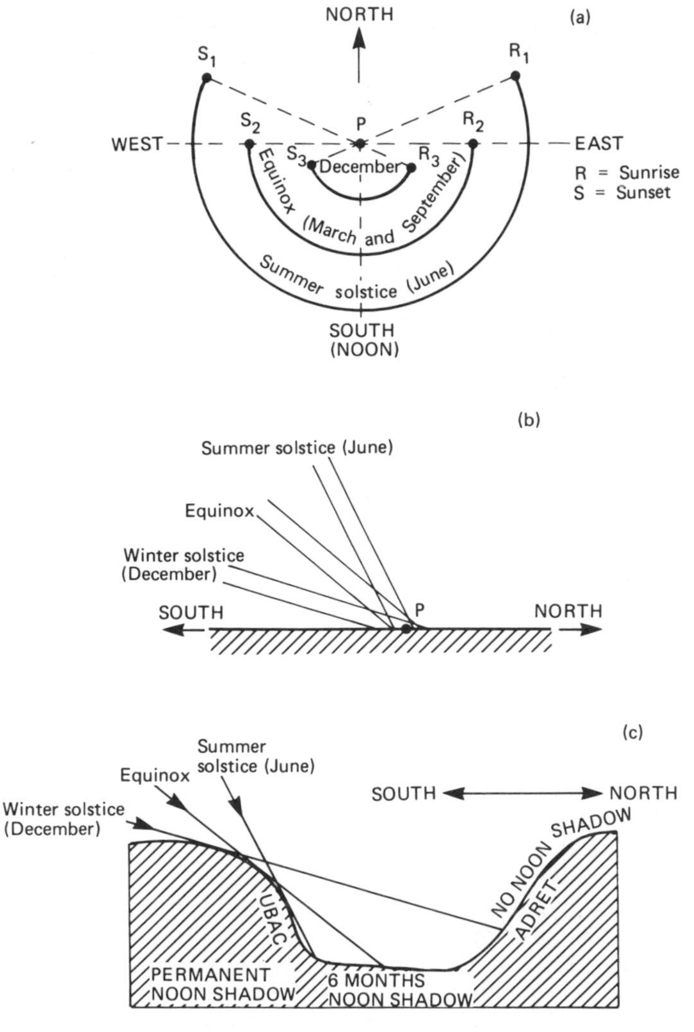

Fig. 2.9 Variations in insolation in relation to season and relief of the land surface

human settlement. It should be noted, however, that in summer the ubac slope can receive some direct sunshine for a period following sunrise and just before sunset (Fig. 2.9(*a*)).

Local climates also differ according to the type of vegetation that covers the ground surface. Forests, particularly coniferous and tropical rain forests have lower albedos than, for example, grassland. Hence, they can absorb a greater proportion of incoming solar radiation and can warm the air above them by radiation of heat. Given the same amount of insolation, therefore, one would expect the air above a forest to be warmer than that above an area of grassland or crops. But temperatures vary within these forests. The canopies of coniferous and tropical rain forests are particularly dense. As a result, most of the absorption and radiation, and consequently the temperature variations, take place at the level of the canopy. Beneath the canopy the temperature remains much more constant since little solar radiation penetrates and the loss of heat by radiation is restricted.

On a rather larger scale than the above, examples of local climates are the temperature differences between cities and their surrounding rural areas. Urban areas are frequently at least 6°C warmer than the nearby countryside. This phenomenon is referred to as a town's 'heat island' and it is particularly noticeable during a spell of calm weather. During the daytime, solar radiation is absorbed by buildings, pavements and roads. Some buildings may even be warmed by sunlight passing through windows. Also, in streets, vertical walls can reflect insolation to other buildings and also to roads and pavements. Central heating systems and heat-using industrial undertakings can also contribute to the rise in temperature. Anticyclonic conditions with an inversion in the lower atmosphere can prevent the loss of heat from the city by convection. Most city surfaces are quite dry, unlike the countryside, so little heat is lost through the evaporation of moisture. At night, the heat that has been stored in the city's buildings and streets tends to be lost by radiation much more slowly than in the countryside, and many sources of urban heat continue to operate through the night. Some cities tend to generate a hazy atmosphere through pollution. This can function in the same way as a cloudy sky and hinder the loss of heat by radiation at night. The highest temperatures in urban heat islands tend to occur in areas with the highest density of buildings. Parks and other open spaces have lower temperatures.

2.2 Atmospheric moisture

Much moisture exists in the atmosphere in the form of a gas, water vapour. Water vapour can be changed to the liquid form (water) by *condensation* or to the solid form (ice) by *sublimation*. Melting and evaporation can cause water and ice to return to the form of water vapour.

CONDENSATION

At any temperature a certain maximum amount of water vapour can be held in the air when it is fully saturated. Usually the air is not completely saturated. Its degree of saturation is measured by its relative humidity. This is a measure of the moisture content of the air expressed as a percentage of the moisture content of the same volume of saturated air at the same temperature. The relative humidity of fully saturated air is 100%. The amount of water vapour that fully saturated air is able to hold is greater if the air is warm than if it is cold (Fig. 2.10). Hence, the relative humidity of the air can rise either by an increase in its content of moisture or by a decrease in its temperature (Fig. 2.10). The temperature at which air becomes fully saturated is known as its dew point. If the air's temperature falls below the dew point, condensation of water vapour to liquid water takes place.

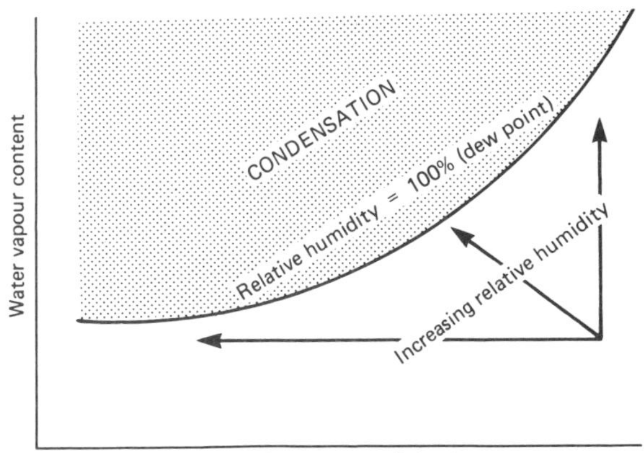

Fig. 2.10 The general relationship between water vapour content and air temperature

Condensation in the atmosphere

One of the main reasons why air may be cooled is that it may rise to a higher altitude. This can occur either as a result of convection or through a stream of air rising either over a mountain range or over another mass of air with different characteristics. For example, warmer, lighter air tends to rise over colder, denser air. As a mass of air rises it becomes cooler because the atmospheric pressure decreases with altitude and this causes the air to expand. Provided that no condensation takes place the air cools at the dry *adiabatic lapse rate* (1°C per 100 m) (Fig. 2.11). Eventually, the rising air reaches the same temperature as the air surrounding it and its upward movement ceases. In the meantime, however, it is possible that the rising mass of air has been cooled to its dew point. If so, it will have begun to cool much more gradually as it rises, at the saturated adiabatic lapse rate. This is a much more gradual rate of decline of temperature (only about 0.6°C per 100 m) (Fig. 2.11) since latent heat becomes sensible heat through the condensation of water vapour. In this case the air will rise to a considerably higher level before its temperature falls to that of its surrounding air.

Three basic conditions of stability in the atmos-

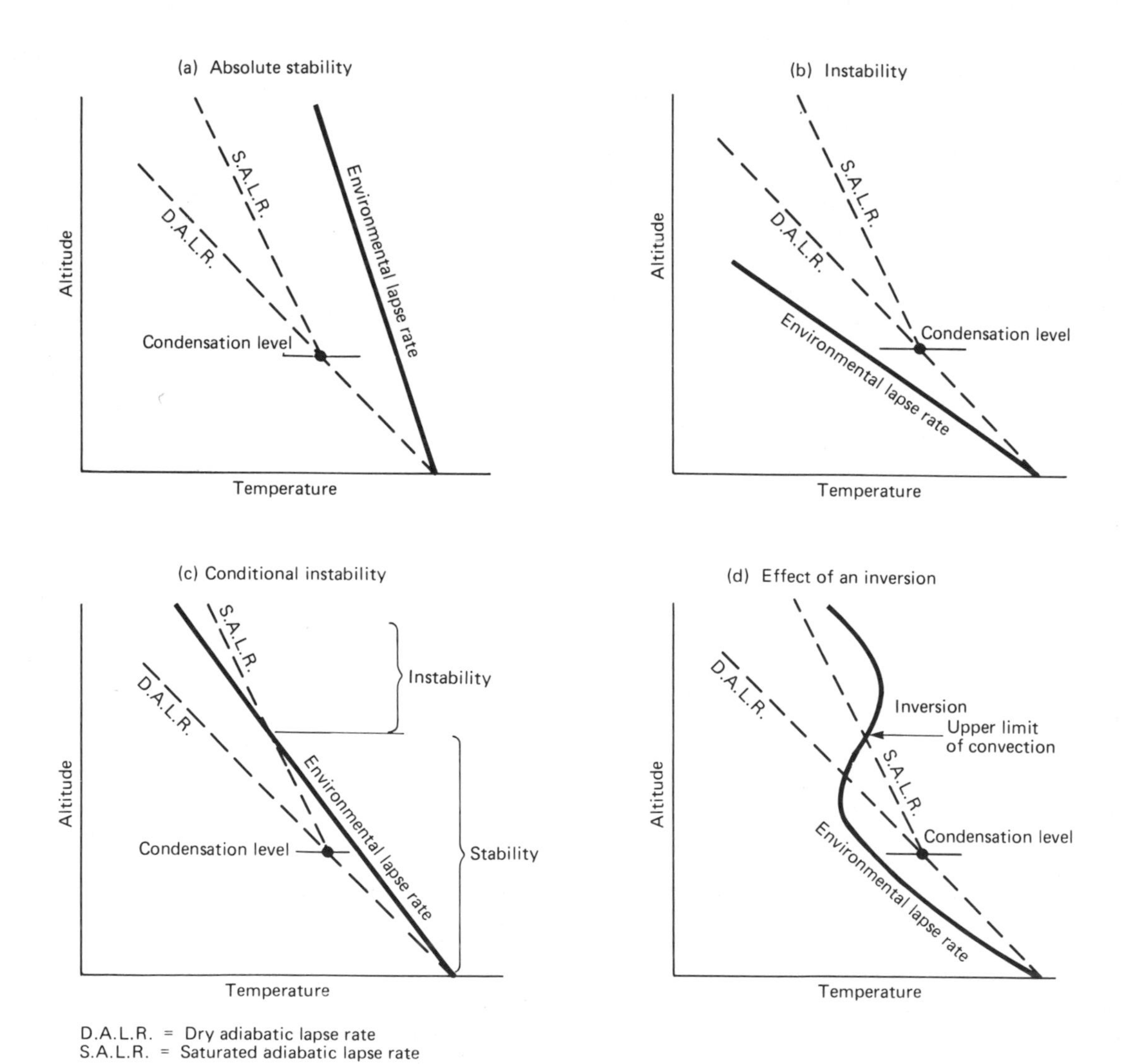

Fig. 2.11 Specimen conditions of atmospheric stability

phere can be recognized. These are related to variations in the *environmental lapse rate* (the average rate at which temperature declines with altitude even if the air is completely still). Absolute stability (Fig. 2.11(*a*)) exists when conditions are such that temperatures in the atmosphere decrease very gradually from the ground upwards. The environmental lapse rate is lower than either the dry or the saturated adiabatic lapse rate. In this case air near the ground surface cannot rise by convection because if it were to rise at all it would become cooler than its surrounding air and it would sink back to the ground surface. Air is said to be unstable if the environmental lapse rate is greater than the dry adiabatic lapse rate (Fig. 2.11(*b*)). This results in vigorous convection since, as soon as ground surface air begins to rise it cools less quickly and so becomes warmer than its surrounding air. When it reaches the condensation level it cools at an even slower rate so convectional uplift is intensified. A third basic condition is that of conditional instability (Fig. 2.11(*c*)). In this case the air is stable at the lowest levels of the atmosphere since the environmental lapse rate is lower than the dry adiabatic lapse rate. If, however, the air stream passes over a range of hills and condensation begins, the rising air cools at the saturated adiabatic lapse rate which may be less than the environmental lapse rate. Provided that there is sufficient uplift the rising air may come to have a higher temperature than the air that it is invading. Hence it becomes unstable. It can be seen therefore that stability exists when the environmental lapse rate is smaller than either of the adiabatic lapse rates (Fig. 2.11(*a*)). Instability exists when the environmental lapse rate is greater than either of the adiabatic lapse rates (Fig. 2.11(*b*)). Conditional instability exists when the environmental lapse rate is intermediate between the dry and the saturated adiabatic lapse rates (Fig. 2.11(*c*)). It is possible for conditions of both stability and instability to exist at the same time. Figure 2.11(*d*) shows a case in which the air in the lowest part of the atmosphere is unstable. Higher up, however, there is an inversion (the air's temperature increases with an increase in altitude). This inversion sets an upper limit to convection.

Condensation in the atmosphere results in the formation of clouds which consist of tiny drops of water or, in some cases, ice crystals. Condensation does not occur easily in clean air. Moisture condenses more readily on tiny particles of dust, smoke, pollen, fine soil or even salt. In general, clouds take the form of either widespread level sheets that may cover the whole of the sky (stratus), or masses that are relatively small when seen from below, but which can extend to very great heights (cumulus). These forms can occur very near the ground (less than 2000 m), at medium heights (2000–6000 m) and at heights above 6000 m. The very high clouds are subdivided into *cirrus* (thin, fibrous streaks), *cirro-cumulus* (sometimes called 'mackerel sky') and *cirro-stratus* (a more complete coverage of the sky through which the sun or moon can be seen, often with a halo). The clouds at medium height are subdivided similarly into *alto-cumulus* (small rounded clouds with clear sky between) and *alto-stratus* (a sheet cloud rather denser than cirro-stratus). Low clouds include *stratus* (a uniform dark grey coverage of the sky), *cumulus* ('lumpy' clouds with a flat base) and *strato-cumulus* (often long rolls of cloud with clear sky between). One type of cloud, *cumulonimbus*, associated with thunderstorms, has an enormous vertical range and can stretch from low levels to well over 10 000 m. It tends to have a flat base at the condensation level and, at its top, rounded projections indicate strongly rising air currents, stimulated by the release of latent heat during condensation. The tops of these clouds sometimes spread laterally to form the shape of an anvil. Because they have such a great vertical development, they exclude sunlight almost completely. Hence, viewed from below, they often look black.

Condensation at or near ground level

When a warm, moist air stream passes over a cold land surface water vapour may be cooled to its dew point and condensation may take place. At temperatures above freezing point dew may be deposited, especially on grass surfaces. Blades of grass have a very large surface area in relation to their volume and on a clear night they lose heat rapidly by radiation. The air near the grass is cooled to its dew point and *dew* is deposited. If ground temperatures are below freezing point *hoar frost* (ground frost) may form instead of dew. Air trapped between the ice crystals gives hoar frost its white colour.

A cold land surface can also influence the temperature of the lower layers of the air during a cold, calm night. The air is cooled by contact with the cold ground and drifts along the ground to accumulate in hollows. Condensation of water

Fair-weather cumulus clouds viewed from an altitude of 10 000 metres over the Atlantic to the south-west of Portugal

vapour within this cold air can produce *radiation* (inversion) *fog*. This type of fog is liable to occur in urban areas where atmospheric pollution provides hygroscopic nuclei in the form of smoke or sulphur dioxide.

Another type of fog is *advection fog* which often occurs when an air stream moves from a warmer to a colder area. This type of fog occurs in Newfoundland where relatively warm air that has been flowing over the North Atlantic Drift reaches the cold waters of the Labrador Current (Fig. 2.8). Also, some of the coastal areas in the subtropics which have cool ocean currents flowing equatorwards (Fig. 2.8) tend to experience mist or fog, especially in the summer season.

PRECIPITATION

Rapid condensation of water vapour in clouds gives rise to forms of precipitation such as rain, hail, snow and sleet which fall to the earth's surface.

This condensation is the result of the different processes whereby air containing water vapour rises to a higher level of the atmosphere and therefore is cooled so that its relative humidity gradually increases. Eventually the dew point is reached.

Condensation can occur as a result of updrafts of vertical convection currents in clouds (page 104, Fig. 2.11). Alternatively, a stream of air may rise steadily over a mountain range (Fig. 2.12(*a*)). As it ascends the windward side it cools at first at the dry adiabatic rate (1°C per 100 m). If it cools sufficiently, condensation may begin and this causes a reduction in the rate of cooling to the saturated adiabatic rate (about 0.6°C per 100 m). When the air stream begins to descend the lee side of the mountains its temperature increases at the dry adiabatic rate. The air becomes drier and considerably warmer (Fig. 2.12(*a*)). This area is said to lie in the rain shadow of the mountains. Such areas frequently experience very warm, dry winds. Examples are the föhn wind in the Alps and the chinook in the Rockies of western Canada. Another possibility is that, as air rises over a mountain range, if the air is conditionally unstable (Fig. 2.11(*c*)), it may reach its condensation level as it rises up the windward slope, thus developing a condition of instability so that it continues to rise above the mountain's summit and causes heavy precipitation on the mountain. A stream of air can also be uplifted if it encounters a cooler mass of air, as at the warm front of a depression (Fig. 2.12(*b*)) or if it is forced upwards as it is undercut by colder air at the cold front of a depression (Fig. 2.12(*b*)). In the centre of a depression, air may rise by a spiral motion. At the cold front of a depression the undercutting of warm air by colder air often produces heavy convectional rainstorms associated with cumulonimbus clouds. The warm front usually has a broader area of precipitation. Cirrus clouds usually occur at a high level as the warm front approaches. As the height of the warm front gradually decreases with the approach of the depression, the clouds gradually descend through alto-stratus to low level stratus which commonly yield rain (nimbo-stratus).

Rain is caused by the coalescence within clouds of droplets of water so that they form drops which are large enough to fall towards the ground. As they fall they increase in size through colliding with other droplets. In some cases ice crystals in clouds combine to form snowflakes which then melt as

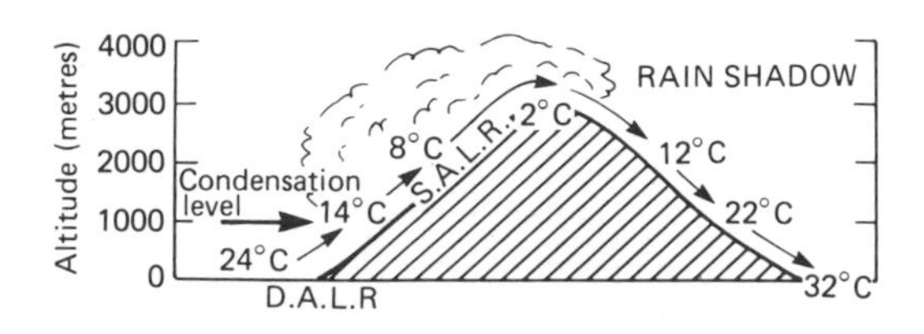

(b) A depression

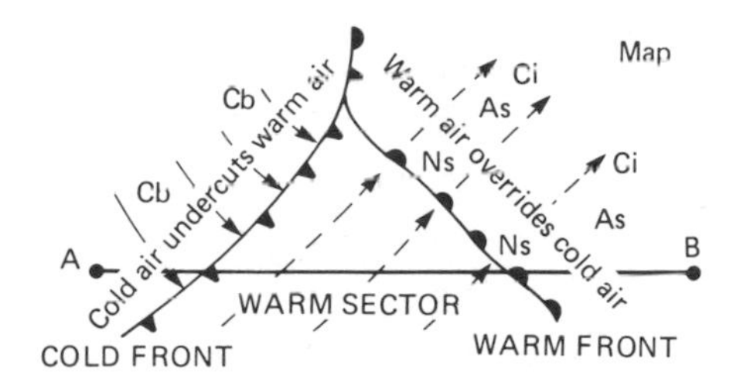

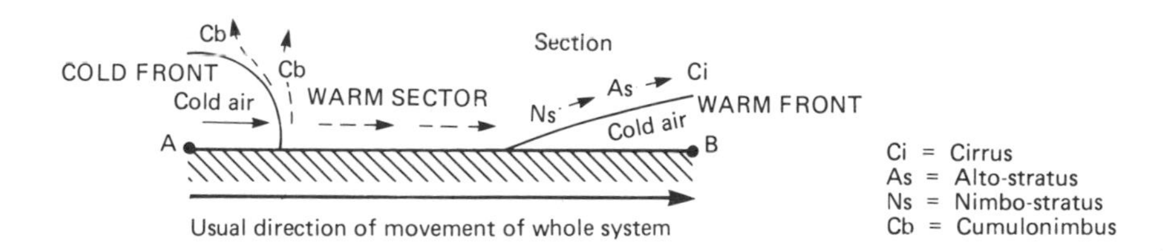

Ci = Cirrus
As = Alto-stratus
Ns = Nimbo-stratus
Cb = Cumulonimbus

Fig. 2.12 Conditions in which precipitation occurs

they fall through the cloud to form raindrops. This latter process tends to produce heavy rain showers. Smaller cloud droplets fall as drizzle. The technique of artificial rainmaking is based upon the 'seeding' of clouds with nuclei (silver iodide or 'dry ice' (CO_2)) so as to encourage the development of ice crystals.

Snow is formed when water vapour in a cloud changes directly to ice. The ice crystals then combine together to form snowflakes. For snow to reach the ground the freezing level in the cloud must be near enough to the ground surface to prevent the flakes from melting completely as they fall. If snow partly melts and reaches the ground as a mixture of rain and snow it is referred to as sleet.

Hail consists of rounded pieces of ice, usually not more than 5 cm in diameter. It is produced in cumulus or cumulonimbus clouds. Drops of rain are carried upwards in the cloud and they freeze. They may then move a number of times upwards and downwards within the cloud collecting a number of extra layers of ice through colliding with supercooled water droplets which freeze on contact. The internal structure of a hailstone therefore shows a number of concentric layers.

2.3 Atmospheric circulation

GENERAL CIRCULATION

Atmospheric systems

Although the solar energy system is in equilibrium when considered at the scale of the whole earth (Fig. 2.1), it is not in equilibrium in all parts of the earth. In general, the input of solar energy decreases from the tropics to the poles mainly because the angle of elevation of the sun decreases towards the poles. Differences in insolation also arise out of variations in cloudiness. Temperature is basically determined by the balance between incoming and outgoing radiation. In general, between latitudes 38°N and 38°S there is a surplus of solar energy (Fig. 2.13(*a*)). The amount of incoming solar radiation is greater than the amount of outgoing terrestrial radiation. One would therefore expect this zone to become hotter and hotter. Similarly, the deficit in respect of solar radiation poleward of latitude 38° might be expected to result in these areas becoming colder and colder. In fact, of course, average temperatures over the world remain roughly constant year after year. This stability is achieved by the transfer of heat (both sensible heat and latent heat) from low to high latitudes, mainly by winds (particularly the westerlies) but also by ocean currents (Fig. 2.7 and Fig. 2.8).

Similarly the hydrological cycle is in equilibrium when considered at the global scale. Total evaporation is equal to total precipitation over the whole earth but this does not apply in different parts of the earth. Evaporation greatly exceeds precipitation between latitudes 10° and 40° north and south where large desert areas exist in the northern hemisphere and large oceans in the southern hemisphere (Fig. 2.13(*b*)). Precipitation exceeds evaporation near the equator and poleward of latitude 40°. This is possible because the westerlies and the trade winds carry moisture polewards and equatorwards respectively. The westerlies of the southern hemisphere are particularly important in this respect (Fig. 2.13(*b*)). It is therefore clear that the solar energy system (Fig. 2.13(*a*)) provides the energy for the operation of the hydrological cycle.

Another aspect of the hydrological cycle is concerned with the exchanges of moisture between the earth's land areas, its oceans and its atmosphere (Fig. 2.13(*c*)). Taken as a whole, the earth's atmosphere, together with its land and sea areas, constitute a closed system. The whole of the earth's water is contained within the system. There are no inputs or outputs. Each of the separate atmospheric, oceanic and land subsystems, however, is an open system, since each both imports and exports water. 77 % of the water precipitated from the atmosphere falls on the oceans and 23 % falls on the land. In return, the atmosphere receives 84 % of total precipitation as evaporation from the oceans and only 16 % from the land areas. The atmosphere, therefore, is in equilibrium. It imports exactly the same amount of water, by evaporation and evapotranspiration, as it exports ($84+16 = 77+23 = 100$). The land areas, on the other hand, have a moisture surplus since they receive 23 % of total precipitation and lose only 16 % by evapotranspiration. The oceans have a moisture deficit. They lose 84 % by evaporation and receive only 77 % by precipitation. Hence, the surplus received by the land areas is equal to the oceans' deficit. The surplus from the land areas, however, flows over the land surface into the

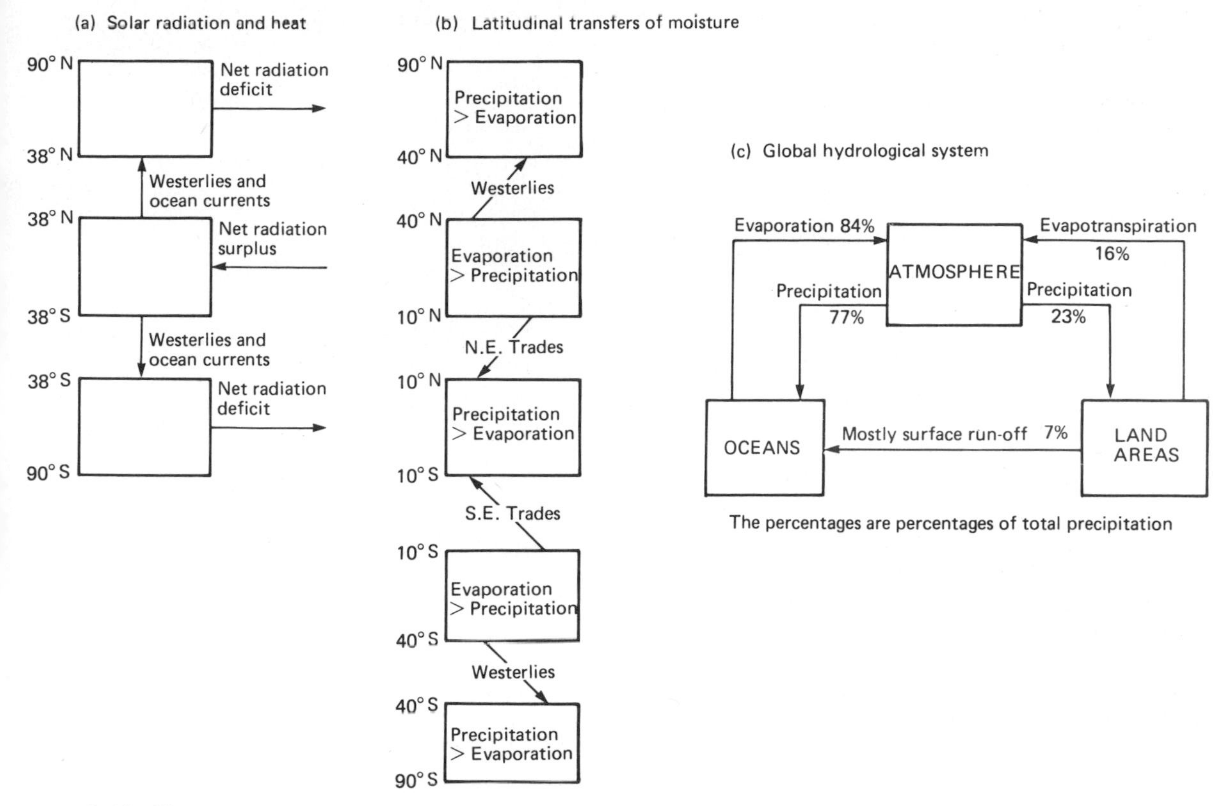

Fig. 2.13 Climatic systems

oceans, mostly as stream flow, and restores the balance (Fig. 2.13(*c*)). Each of the subsystems of land, ocean and atmosphere has been described above in terms of what has been called the 'black box' approach, in which the subsystem has been treated as a unit, and no attention has been paid to smaller subsystems that exist within it. The 'land areas' subsystem, of course, contains many other interlinked subsystems of smaller scale, such as hillside slopes (Fig. 1.13), drainage basins (Figs. 1.14 and 1.19), river long profiles (page 43) and the solar energy system (Fig. 2.1).

General model of atmospheric circulation

Figure 2.14 illustrates the main characteristics of circulation in the troposphere. Near the equator insolation is at a maximum and temperatures are high. Air at the surface tends to rise by convection. Near the tropopause this air tends to move northwards and southwards and then to descend at about latitude 30° north and south, at the subtropical high-pressure belts. As it descends it warms adiabatically (page 106). Hence, the subtropical high pressure belts are areas of atmospheric stability with light winds and clear skies. Many of the world's areas of desert are found here. High above these areas, in the upper troposphere, are the subtropical jet streams, belts of very strong winds from which air also subsides. At the surface, this descending air moves both polewards as the westerlies (south-west in the northern hemisphere, north-west in the southern) and equatorwards as the trade winds (north-east in the northern hemisphere and south-east in the southern). The trade winds flow along the pressure gradient and converge on the low pressure area near the equator. A simple system of circulation such as this is known as a Hadley cell. The trade winds tend to have a constant direction and strength and are often capped by an inversion that prevents convection and results in a generally dry climate. They converge in the *Intertropical Convergence Zone* (ITCZ) and give considerable cloud and rain, especially in land areas. They change their latitude very little through the year, except in the Indian Ocean.

Atmospheric stability, with low stratus clouds, at Tenerife (latitude 28° N)

Air also moves polewards from the subtropical high pressure belts in the form of the mid-latitude westerlies, generally regarded as blowing from the south-west in the northern hemisphere and from the north-west in the southern hemisphere. These are quite unlike the trade winds in that their direction and strength vary greatly from time to time. This is because they are influenced from time to time by depressions (centres of low pressure) into which air converges, and anticyclones (centres of high pressure) from which air diverges. These result in rotary movements of air in an approximately horizontal plane. This happens particularly in the northern hemisphere. In the southern hemisphere the westerlies are stronger and more constant in direction. In the polar areas there is general atmospheric subsidence and air moves generally equatorwards (polar easterlies), towards the sub-polar low pressure area which, rather than being a continuous latitudinal belt, reflects the passage of a series of depressions in which air is circulating round a centre of low pressure and winds are therefore variable in direction. Just below the tropopause, at the polar front, is the polar front jet stream, in a position where there is a steep temperature gradient between polar air and tropical air (Fig. 2.14).

Because of the tilt of the earth's axis, these wind and pressure belts tend to migrate through several degrees of latitude according to the season of the year, northwards in the northern summer and southwards in the southern summer.

Major pressure and wind systems

The major pressure and wind systems over the earth (Fig. 2.15) have a general resemblance to those illustrated in Figure 2.14 but there are certain differences which appear to be related to the distribution of land and sea over the earth's surface. For example, land areas in the northern hemisphere in January (winter) tend to develop high pressure (Fig. 2.15(*a*)) partly because of their

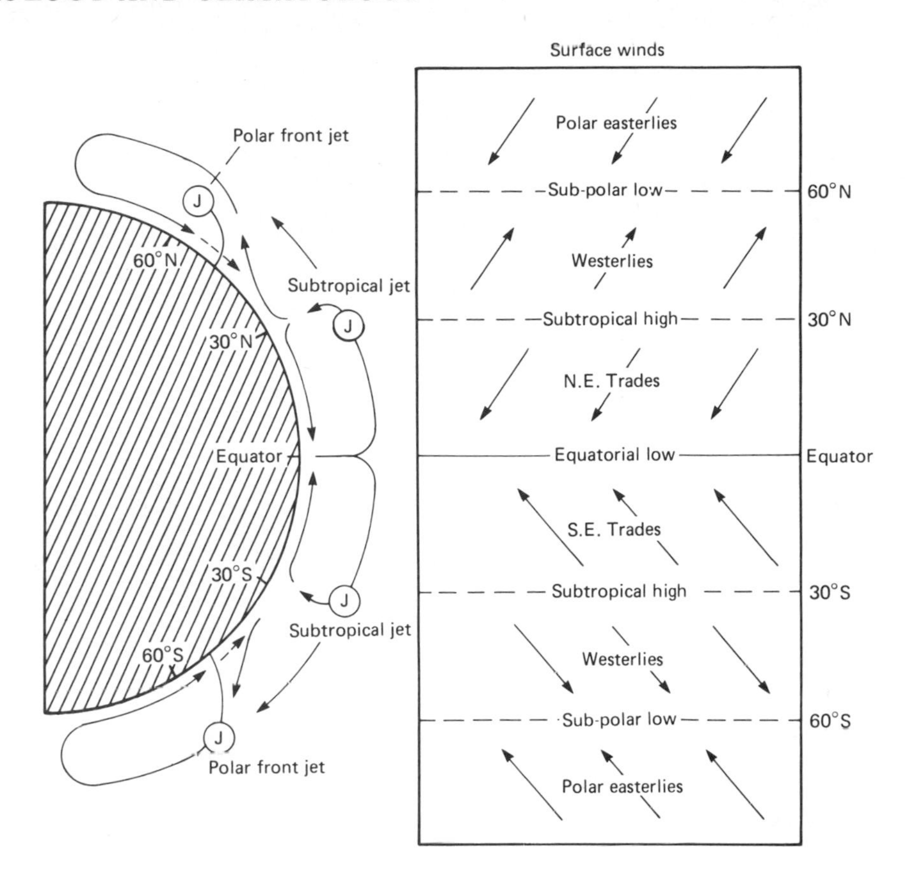

Fig. 2.14 Atmospheric circulation (simplified)

low temperatures. Thus, in the northern winter, the sub-polar low tends to be centred over the north Atlantic, near Iceland and over the north Pacific, near the Aleutian Islands. In summer (July) (Fig. 2.15(*b*)), the Icelandic low is still prominent but relatively low pressure extends across the northern continents. In contrast, in the southern hemisphere, the sub-polar low is a fairly regular, latitudinal belt.

In theory the distribution of the subtropical high pressure belts should resemble the model more closely. Their location is not so much the result of temperature differences as of the dynamic forces of the circulation of the atmosphere (Fig. 2.14). To the west of longitude 60°E they do occupy approximately their expected locations at about latitude 30° degrees north and south, beneath the subtropical jet streams. In the northern hemisphere in January, however, they tend to intensify over the continents, particularly North America. In July they are most intense over the oceans, in the Atlantic between Bermuda and the Azores, and in the Pacific centred near the Hawaiian Islands (Fig. 2.15(*b*)). In the southern hemisphere the subtropical highs are best developed over the oceans and they move slightly polewards in summer (January) and equatorwards in winter (July). Relatively low pressure exists along the equator at all seasons.

The subtropical high pressure belts have a very strong influence on the location of both the trade winds and the westerlies. Both sets of trade winds (north-east and south-east) blow generally towards the equator and they maintain a fairly constant direction and strength. Near the equator the two sets of trade winds converge in a zone of generally rising air (Fig. 2.14) and low pressure known as the Intertropical Convergence Zone (ITCZ). Here, there is a considerable amount of cloud and rain, especially over land areas.

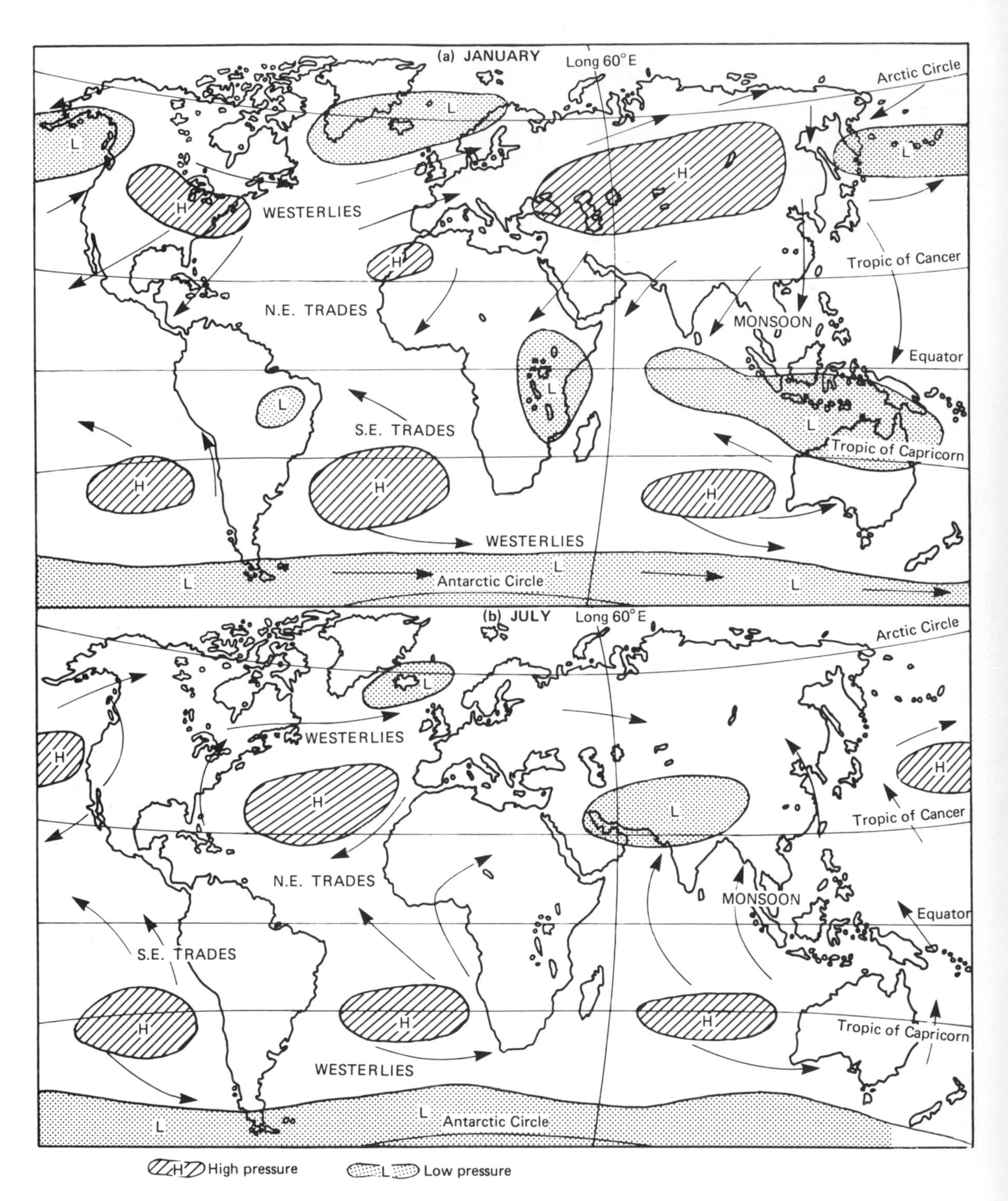

Fig. 2.15 Major pressure and wind systems at the earth's surface

The subtropical high pressure belts are also the origin of the mid-latitude westerly winds. In the northern hemisphere the westerlies, unlike the trade winds, are not a system of winds blowing from a particular direction. They consist rather of areas of low pressure and high pressure that move generally north-eastwards. Winds are therefore variable both in direction and velocity. In general, because of the progressive increase in the strength of the Coriolis force with increasing latitude, the westerlies tend to swing progressively towards the east as they move northwards, and their velocity tends to increase. The polar easterlies (Fig. 2.14) occur chiefly on the poleward sides of low pressure cells that are moving predominantly eastwards. In the southern hemisphere the westerlies are generally stronger and more constant in their direction.

In south-east Asia, east of longitude 60° E, there is a different distribution of land and sea. Instead of there being north-south trending continents (the Americas and Europe and Africa) and oceans (the Atlantic and the Pacific) there is a large continent (Asia) in the north and an ocean (the Indian Ocean) to the south. The main effect of this is that in the northern summer (July) (Fig. 2.15(*b*)) northward moving winds replace the normal north-east trades in southern Asia. In the southern hemisphere these winds resemble the south-east trades but, as they cross the equator, the earth's Coriolis force tends to swing them towards the north-east, so that they become the south-west monsoon. It seems clear that this flow of air is related to the development of low pressure in the northern summer over southern Asia (Fig. 2.15(*b*)), particularly to the south of the Tibet Plateau. This may be due to the fact that the heating effect of solar radiation is greater over the land area than in the ocean to the south, since, over the ocean, much heat energy is used for evaporation. Alternatively it has been suggested that, in this region, in the northern summer, the Intertropical Convergence Zone shifts further to the north than it does elsewhere. To some extent, a similar monsoon system exists in northern Africa (on the west coast) where air from the south Atlantic invades the westward 'bulge' of Africa. In this area the ITCZ reaches as far as latitude 20° N in the northern summer (Fig. 2.15(*b*)), allowing south-west winds to enter.

AIR MASSES

Air masses are large volumes of air which are more or less uniform in respect of temperature and humidity over very large areas. The environmental lapse rate of temperature tends to be uniform over the whole of a particular air mass. An air mass gains these uniform characteristics through remaining for a long period in a 'source region'. Here, it is influenced by the particular temperature and humidity characteristics of the land or sea surface with which it is in contact.

Air masses are classified according to the characteristics of their source areas. 'Arctic' and 'polar' air masses originate in high latitudes, generally polewards of latitude 50 degrees north and south. They can be further subdivided into 'continental' or 'maritime' according to whether they are created over a continent or an ocean. Hence, for example, the continent of Antarctica is a source of 'continental arctic' (cA) air and the northern parts of North America and Asia are source regions for 'continental polar' (cP) air (Fig. 2.16). Most of the ocean surrounding Antarctica is a source of 'maritime polar' (mP) air, as are the northern parts of the Atlantic and Pacific Oceans. 'Tropical' air masses are created in the subtropical high pressure systems at about latitude 30° north and south. 'Maritime tropical' air masses originate over the oceans in these latitudes and 'continental tropical' air masses originate mainly over the larger land areas such as southern Asia and northern Africa (Fig. 2.16). The most influential of these air masses are the ones that originate in the subtropical high pressure belts of the Atlantic and the Pacific Oceans. Air masses move from their source areas into zones of convergence which are areas of relatively low pressure. These areas may be either warmer or colder than the source areas. Hence, the lower layers of the moving air mass may be modified by contact with the underlying land or sea surface. This possibility is catered for in the air mass classification system by the use of the suffixes 'w' and 'k'. The suffix 'w' indicates that the air mass is warmer than the surface over which it is passing; 'k' indicates that it is colder. Air masses of the 'w' type tend to be cooled in their lower layers, thus producing an inversion (Fig. 2.11(*d*)). Thus, convection is unlikely to occur and the air mass is comparatively stable. Such conditions can occur where a warm moist mT (maritime tropical) air mass passes over a relatively cool sea where there is a cool ocean current (Fig. 2.8). In this case, fog or low level stratus clouds can form. Coastal fog often occurs in south-west England in spring with the approach of a maritime tropical air mass over a sea surface whose temperature is at practically the lowest level that it reaches in the whole year. Coastal fog off Newfoundland can occur for similar reasons with a warm air mass passing over a cold ocean current (Fig. 2.8). On the other hand, air masses of the 'k' type become warmer in their lower layers. This tends to steepen the environmental lapse rate (Fig. 2.11(*b*)) and thus cause instability. This can lead to convection currents and the development of clouds of the cumulus type (page 105).

Figure 2.16 shows the approximate locations of the source areas of the major air masses that affect the weather of Eurasia and North America in winter. In summer these source areas tend to shift northwards. A cP air mass is created in winter over the very cold northern parts of Europe and Asia. This is a stable air mass with a layer of very cold, dry air near the surface capped by much warmer air, thus creating an inversion. As it passes over land areas it retains these characteristics of coldness and dryness. If it passes over a sea area, such as the North Sea, whose surface is relatively warm, its lowest layers are warmed (cPk). Low-level clouds may form but their vertical development is limited by the inversion. Maritime polar air (mP) usually approaches Western Europe from the Atlantic ocean. It tends to be unstable (mPk) if it approaches from the north-west and can give heavy showers. If it takes a more southerly track and then swings northwards towards Britain in winter it may acquire mPw characteristics as its

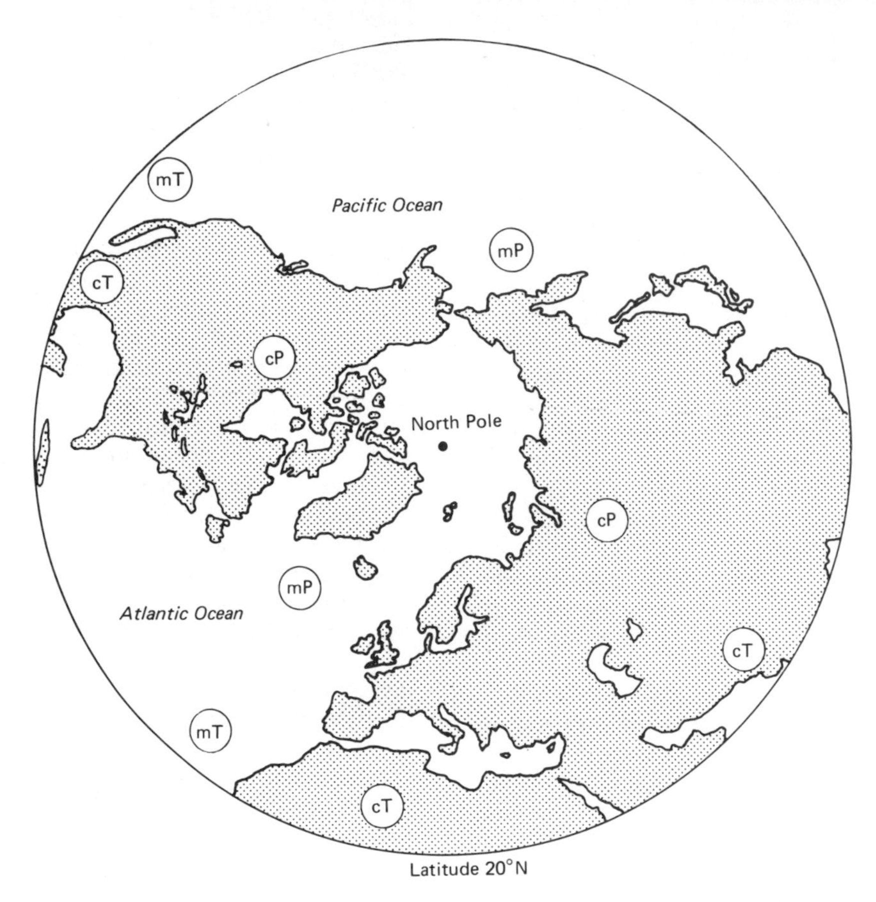

Fig. 2.16 Source areas of some of the air masses of the northern hemisphere in winter

An approaching warm front

lowest layers are cooled, thus forming an inversion. Beneath this inversion stratus clouds may form and give mild, dull, drizzly weather. Maritime tropical air, on the other hand, is very much warmer in its source area which is well to the south. As it moves northwards it tends to be cooled in its lower layers (mTw). This often gives stratus clouds and light rain or drizzle beneath the inversion. Similar relationships to those described above also exist in North America.

Fronts

Fronts are the boundaries between contrasting air masses. An air mass has fairly uniform characteristics over its whole area but these may differ greatly from those of an adjacent air mass. Hence, very sudden changes in, for example, temperature or humidity can occur at a front. Steep temperature gradients across the line of a front can cause strong winds to occur high above the front. A good example of this is the jet stream that occurs above the polar front (Fig. 2.14). A front tends to slope in such a way that, at the front, warmer air overlies colder air (Fig. 2.17(*a*)). The warm air can rise above the wedge of colder air or the colder air can undercut the warmer air and lift it upwards. Both of these processes can result in the formation of clouds and precipitation along the front. The front itself can change its position so that, for example, warm air replaces some of the cooler air (Fig. 2.17(*b*)). In this case, the front is referred to as a warm front. If the cooler air replaces the warmer air the front is a cold front (Fig. 2.17(*c*)).

Depressions

Depressions are systems of low pressure that form along the fronts between contrasting air masses. At first the front may be stationary and fairly straight (Figs. 2.17(*a*)) and 2.18(*a*)). In the northern hemisphere there is south-west flowing cold

Cumulus clouds in a cold northerly air stream over southern England in December. Snow fell in northern England

air on its northern side and north-east flowing warmer air on its south side. At the front the warm air partly overlies the colder air (Fig. 2.17(*a*)). Clouds tend to form in the warm air just above the front and some precipitation may occur. In places, the warm air tends to displace the cold air and produce a shallow salient or wave (Fig. 2.18(*b*)) bounded by a warm front and a cold front. Pressure then begins to fall at the apex where the warm front meets the cold front. This wave travels along the front in the same direction as the warm air is moving. Frequently the warm air penetrates more deeply into the cold air mass and the amplitude of the wave increases (Fig. 2.18(*c*)). A closed centre of low pressure then forms at the junction of the warm front and the cold front. The warm front is now clearly distinguished from the cold front and the 'warm sector' is established between the two fronts. This system then tends to move north-eastwards. At the warm front the warm air is rising over a very gently sloping wedge of colder air usually with very little turbulence.

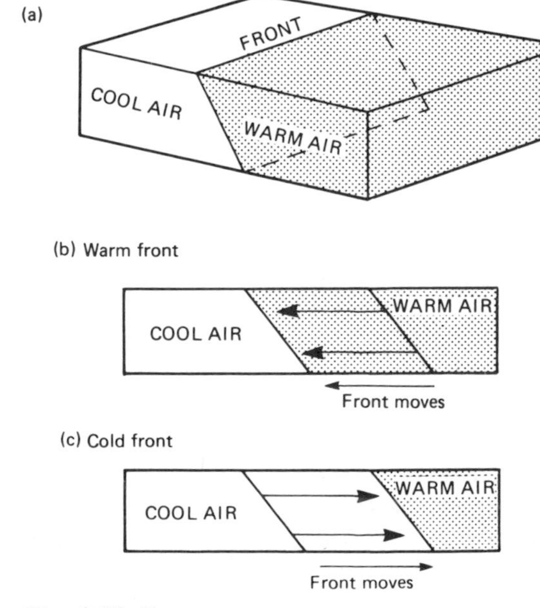

Fig. 2.17 Fronts

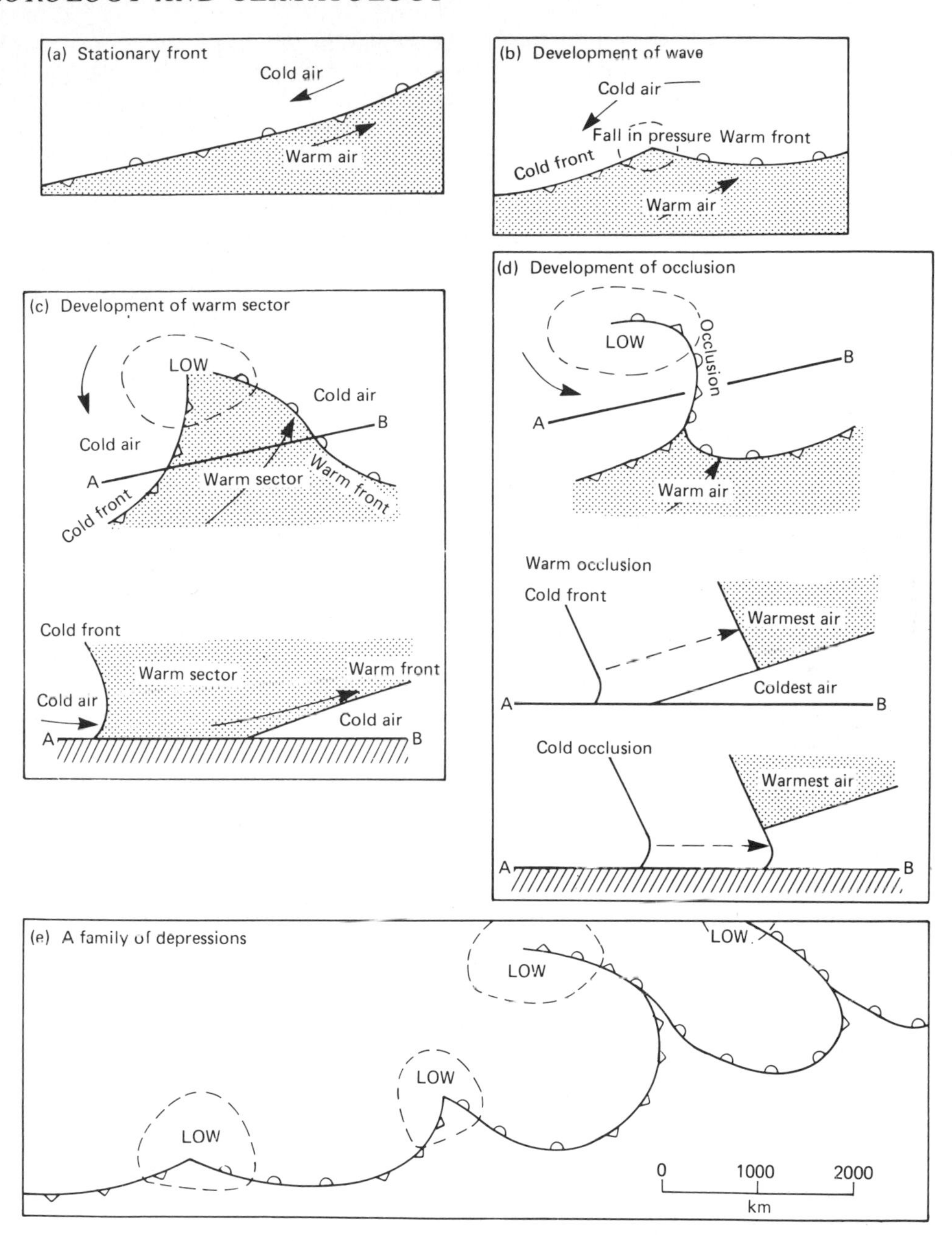

Fig. 2.18 Development of a depression

This tends to produce a wide band of cloud extending some 200 or 300 km in advance of the intersection of the warm front and the surface. High clouds can be seen some twelve hours before the actual arrival of the front at ground level. These are usually of the cirrus type (page 105) and they are succeeded by stratus clouds at gradually decreasing heights as the front approaches. In some cases instability can develop and cumulonimbus clouds can give heavy showers. Various belts of rain have an irregular distribution over the area of the front. These are the characteristics of a warm front of the 'ana' type, in which the air in the warm sector is rising. At a kata-warm front the warm sector air is descending. Hence there are fewer clouds at a high level and only light rain or drizzle tends to occur. As the warm front passes, the wind veers from somewhere near north-east to south or south-west, and the temperature rises. Between the two fronts is the warm sector which has changeable weather with stratus clouds and some drizzle. The cold front is much steeper than the warm front and the colder air tends to undercut the air of the warm sector and displace it fairly violently upwards. At the cold front, therefore, cumulus and cumulonimbus clouds tend to form (page 105) and these give heavier showers than are usual at the warm front or in the warm sector. This is particularly the case at ana-cold fronts. There is also a fairly sudden fall of temperature as the cold front passes and generally brighter weather appears, often with scattered showers.

As time passes, the cold front tends to overtake the warm front and thus the air of the warm sector is lifted to a higher level and the cold air mass completely covers the surface. This is known as an occlusion (Fig. 2.18(*d*)). Sometimes the cold air behind the cold front is warmer than the air in front of the warm front. In this case a warm occlusion is formed (Fig. 2.18(*d*)) with the air from the rear of the depression rising above the air in front of it. The weather in such a system tends to be similar to that experienced at a warm front, with predominantly stratiform clouds. If the cold air to the rear of the depression is colder than that in advance of it a cold occlusion may occur (Fig. 2.18 (*d*)) which provides weather more like that of a cold front, with cumuliform clouds.

A new wave may form on the cold front of a depression and this may develop into a new depression. Thus it is possible for a frontal system to include a number of depressions (Fig. 2.18(*e*)). In this case the ones located furthest to the east have a greater tendency to be occluded than the more recently developed ones.

The development of systems of depressions in the lower troposphere is closely related to conditions in the upper troposphere. Although a depression originates in an area where air masses are converging and air constantly enters the depression, atmospheric pressure is low and may even decrease as the depression develops. For this to occur, *divergence* (a net outflow of air) must remove the rising air in the depression more quickly than low-level *convergence* provides new supplies of air. In the upper troposphere the eastward flow of the upper westerlies and their associated jet stream, in the zone where cold polar air meets warmer tropical air (Fig. 2.14), form a number of large undulations known as *Rossby waves* and also smaller undulations (Fig. 2.19(*a*)). When these undulations extend towards the eq-

(a) Convergence and divergence in the upper troposphere

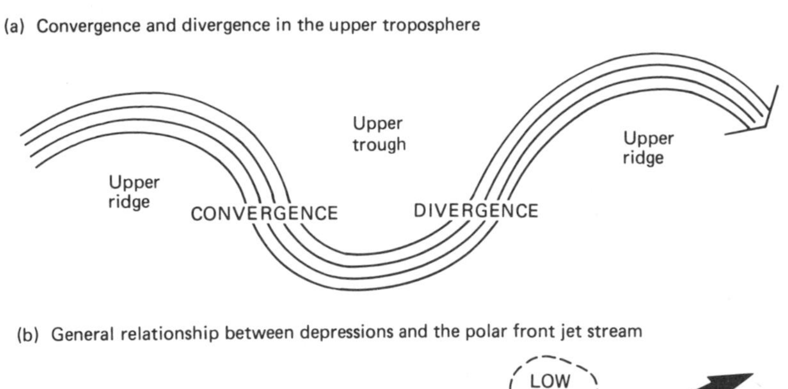

(b) General relationship between depressions and the polar front jet stream

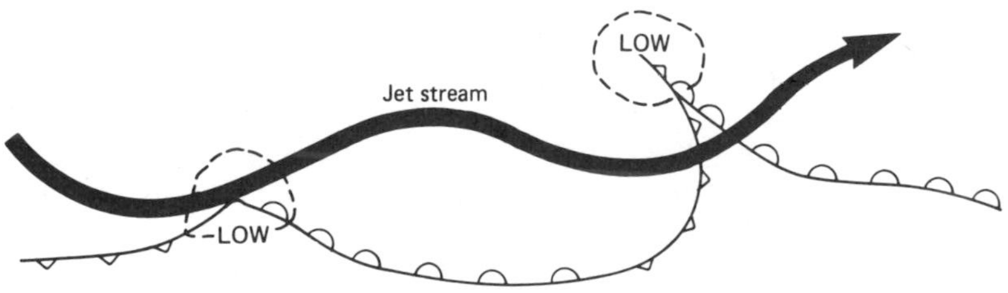

Fig. 2.19 Upper air influences and depressions

uator they form an upper-air trough; when they extend polewards they form a ridge. The effect of these undulations is that the flow of the upper westerlies is alternately convergent and divergent. It is convergent on the west side of a trough and divergent on the east side where its tendency to turn in an anticlockwise (cyclonic) direction is reduced. It is in these areas, to the east of upper-air troughs that low-level depressions form and develop. Their converging air at low levels is dispersed by the divergence in the upper troposphere. Hence, depressions tend to be related to the polar front jet stream as shown in Figure 2.19(*b*). Fairly persistent troughs in the Rossby waves occur to the east of the Rockies of North America and the Tibet Plateau and these can cause the development of depressions.

Anticyclones

Anticyclones are areas of relatively high atmospheric pressure, and pressure increases towards their centres. Their isobaric pattern tends to be circular and they are usually much larger than depressions. In anticyclones the wind rotates in a clockwise direction in the northern hemisphere and anticlockwise in the southern hemisphere. One type of anticyclone occurs in association with depressions. Convergence in the upper westerlies of the northern hemisphere takes place on the east side of an upper air ridge (Fig. 2.19(*a*)). This convergence results in subsidence of air and permits anticyclonic divergence at lower levels. Thus, anticyclones can alternate with depressions in temperate latitudes. Such an anticyclone can grow in size if high level convergence and subsidence deliver more air to it than it loses by low level divergence. Longer lasting anticyclones can develop in temperate latitudes when the curvature of the track of the upper westerlies increases so that the amplitude of their meanders increases and for much of the time they are travelling either north-south or south-north. In these cases longer lasting anticyclones are created which interfere with the free movement of depressions. These so-called 'blocking anticyclones' interrupt the normal sequence of depressions and can cause, for example, unusually cold weather in winter. Cold anticyclones tend to develop on the very cold land surface of Asia and North America in winter. Their coldness may be due to the cooling effect of the underlying land surface and they may be quite shallow. In some cases, however, they are deeper and the upper air is also very cold. Such anticyclones are sources of continental polar air (page 113). In contrast, a warm anticyclone has warmer air at its base. The subtropical high pressure systems (Fig. 2.14) are examples of these. Here, at about latitude 30° north and south there is large-scale atmospheric subsidence and the air is warmed adiabatically as it descends. Thus, there is little possibility of condensation and cloud formation. However, at a height of 2000–3000 m an inversion commonly occurs capping the relatively cooler air near the surface, particularly over the sea. This inversion limits the vertical development of clouds, but stratus clouds can sometimes occur beneath the inversion. In Britain anticyclones tend to give bright sunny days in both summer and winter but at night, particularly in winter, the clear skies encourage rapid radiation of heat and the air closest to the ground is chilled so that condensation takes place which can result in the formation of dew, hoar frost and inversion fog (page 105).

Hurricanes (typhoons, tropical cyclones)

Hurricanes are very deep centres of low pressure that are very much smaller in area than mid-latitude depressions. They appear on weather maps as systems of very closely spaced, circular isobars. They originate in areas of very warm seas fairly near the equator and then they move generally westwards and eventually swing round the western ends of the subtropical anticyclones (Figs. 2.15 and 2.20). The whole hurricane system usually moves at a speed of about 20 km per hour but the winds within the system can flow at speeds of over 120 km per hour. As a hurricane swings polewards it fills rapidly and it may degenerate into a mid-latitude depression as it moves beyond the tropics. Hurricanes occur chiefly on the western sides of the oceans (Fig. 2.20). They are rare near cool ocean currents. Very few occur in the Atlantic Ocean south of the equator. They most commonly occur in late summer or early autumn. A hurricane has a central 'eye' with a diameter of about 30 km which is often calm and comparatively free of cloud. Here, the air is gently descending and warming by compression. This is surrounded by an area of strong winds, thick cumulonimbus clouds and heavy rain. Here, the air is rising strongly. The energy of the hurricane appears to derive from the latent heat released by condensation in this zone. The high wind speeds are partly due to the lack of surface friction over

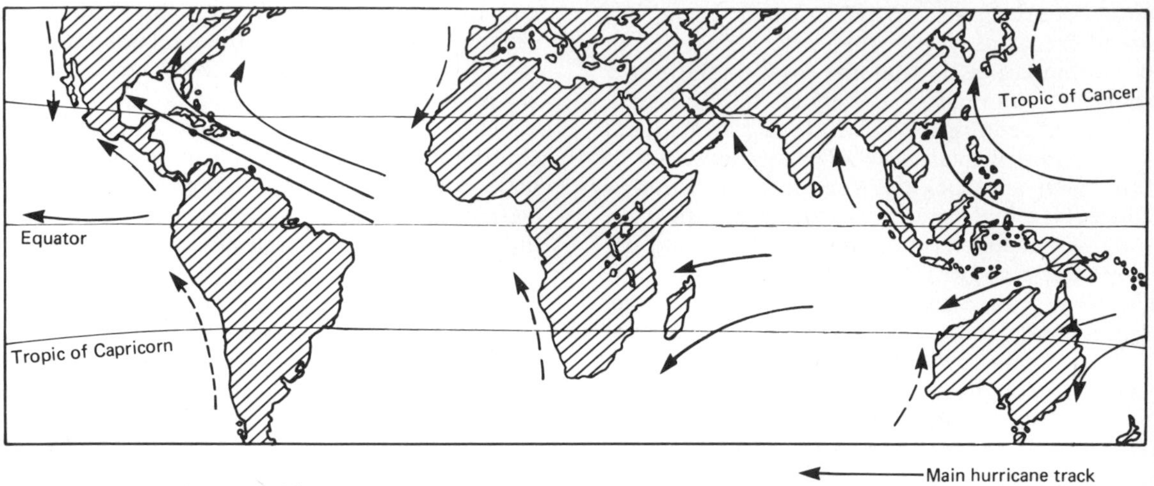

Fig. 2.20 Distribution of hurricanes

the sea. This extreme convergence of air at low levels is balanced by divergence in the upper troposphere, as in the case of depressions (page 119). Hurricanes decay rapidly when they pass over land probably because surface friction causes the rotating air to move inwards and, thus, the centre of low pressure fills.

2.4 World Precipitation

The term 'precipitation' includes rain, snow, sleet and hail, but 'rainfall' is usually regarded as synonymous with precipitation.

Annual precipitation

Rainfall is heavy almost everywhere along the equator, the chief exceptions being the west coast of South America and the east coast of Africa. Much of the rainfall is the result of convection in the Intertropical Convergence Zone (ITCZ). Moist maritime equatorial (mE) air masses give frequent thunderstorms. High rainfall totals also occur along east coasts northwards and southwards from the equator where moist maritime tropical (mT) air flows as the trade winds along the equatorward sides of the subtropical high pressure systems (Fig. 2.15). In south-east Asia, from India to Japan, heavy rain is caused by the monsoon winds blowing from sea to land (Fig. 2.15(*b*)). The south-facing west African coast is also very wet because of a similar, but smaller, monsoon system (page 113, Fig. 2.15(*b*)). In south-east Asia, eastern Australia and central America especially, hurricanes contribute substantially to annual rainfall totals. Areas of relatively high rainfall extend from the tropics along the east coasts of temperate regions. These are caused by the migration of moist maritime tropical air masses into south-east USA, eastern South America and eastern Australia. The remaining areas of high rainfall totals are west coasts in higher latitudes. Here, westerly winds and trains of depressions give high rainfall totals, especially where the coastline is mountainous, as in Alaska, British Columbia, western Scotland, and Norway, and also in Patagonia and New Zealand in the southern hemisphere (Fig. 2.15). The largest arid area of the world extends from west Africa to central Asia. Much of central and western Australia is also arid. Deserts form relatively narrow coastal strips in South America and south-west Africa but, in North America, the desert extends inland into the western cordilleras. Areas in high latitudes of the northern hemisphere also have very low precipitation totals but they also have very low temperatures and they are not usually classified as deserts.

There are three main reasons for the existence of deserts. The deserts that are located near the Tropics of Cancer and Capricorn, and in particular the desert of northern Africa, are strongly influenced by the subsiding air of the subtropical

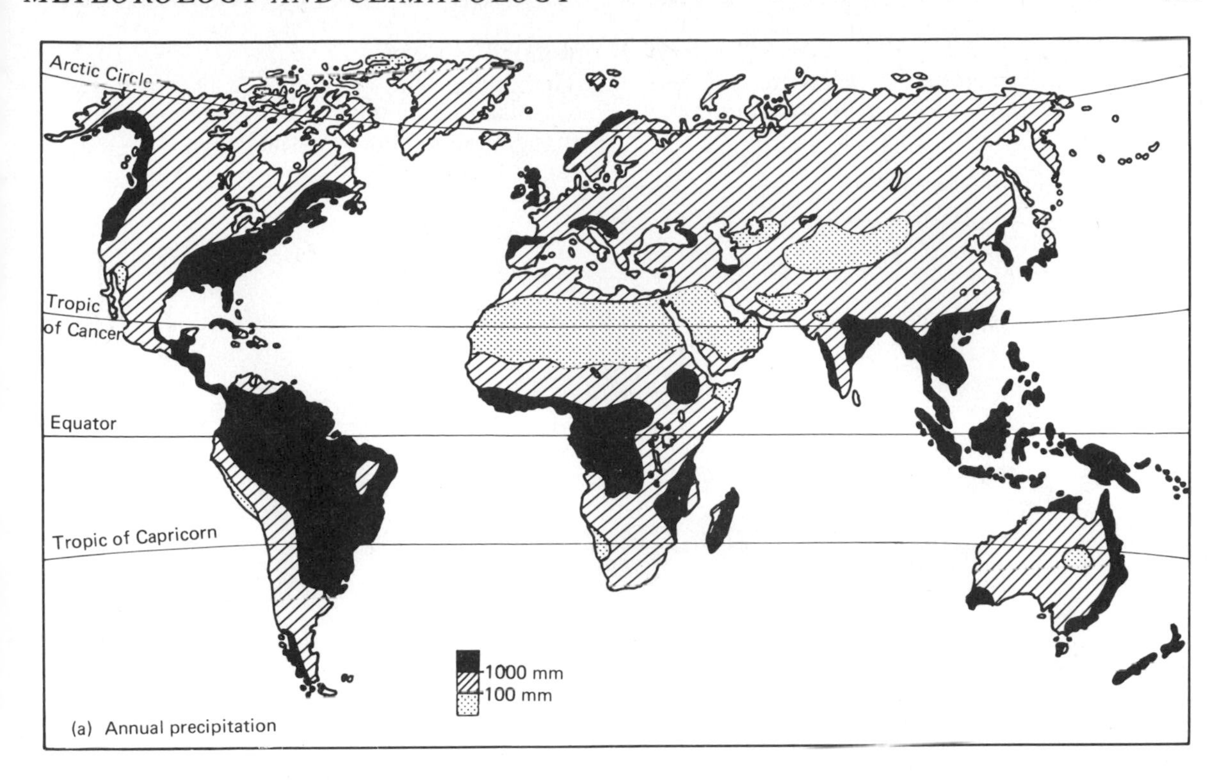

Fig. 2.21(a) World precipitation

high pressure cells (Fig. 2.14). Here, the air is warmed adiabatically as it descends and its relative humidity decreases. Northern Africa is a major source area for continental tropical air (page 114, Fig. 2.16). The deserts of central Asia are covered by an area of high pressure in winter (Fig. 2.15) which tends to deflect maritime air masses approaching from the west. In summer, mountain ranges such as the Himalayas to the south and European mountains to the west tend to obstruct the entry of maritime air masses. The western cordilleras create rain shadows in the west of North America. An outstanding example of a desert of the 'rain shadow' type is Patagonia, at the southern tip of South America, to the east of the Andes. Here the annual rainfall decreases from over 2000 mm in the west to about 120 mm in the east in a distance of about 400 km.

Seasonal variations in precipitation

Figures 2.21(*b*) and (*c*) illustrate variations in precipitation from season to season. These variations are due mainly to the northward and southward migration of the world's pressure and wind systems. In July (midsummer in the northern hemisphere) most of the areas receiving at least 100 mm of rainfall are in the northern hemisphere. They are mostly located between the equator and the Tropic of Cancer; in Africa and the northern part of South America and central America. In south-east Asia they extend north of the Tropic of Cancer to northern Pakistan in the west and Japan and north China in the east (Fig. 2.21(*b*)). At this time the ITCZ is situated north of the equator. In January (midwinter in the northern hemisphere) the tropical and subtropical areas of heavy rainfall shift southwards with the southward migration of the overhead noonday sun and the ITCZ. Areas of heavy rain now cover large areas of South America, southern Africa and north-east Australia (Fig. 2.21(*c*)). In temperate regions in North America and Eurasia more rain falls in midsummer (July) than in January. However, on their west coasts the narrow strips of heavy rainfall extend farther south (to the Mediterranean and southern California) in winter (January). Similarly, in the southern hemisphere, southern Australia is wetter in winter (July) than in summer (January).

Rainfall regimes

The upper half of Figure 2.22 illustrates the seasonal rainfall regimes of a number of locations in the Americas that are situated on the west coast,

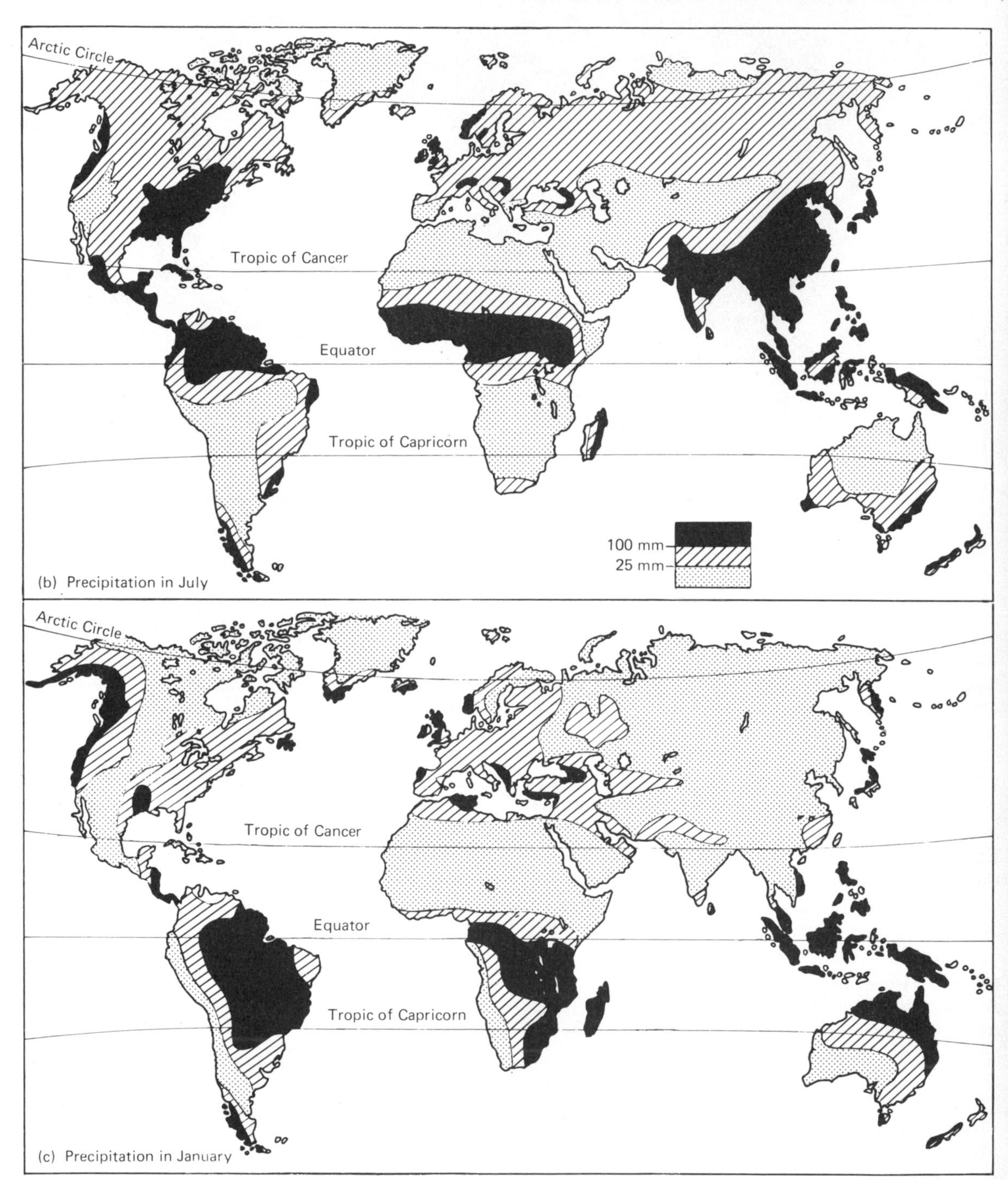

Fig. 2.21(b)(c) World precipitation

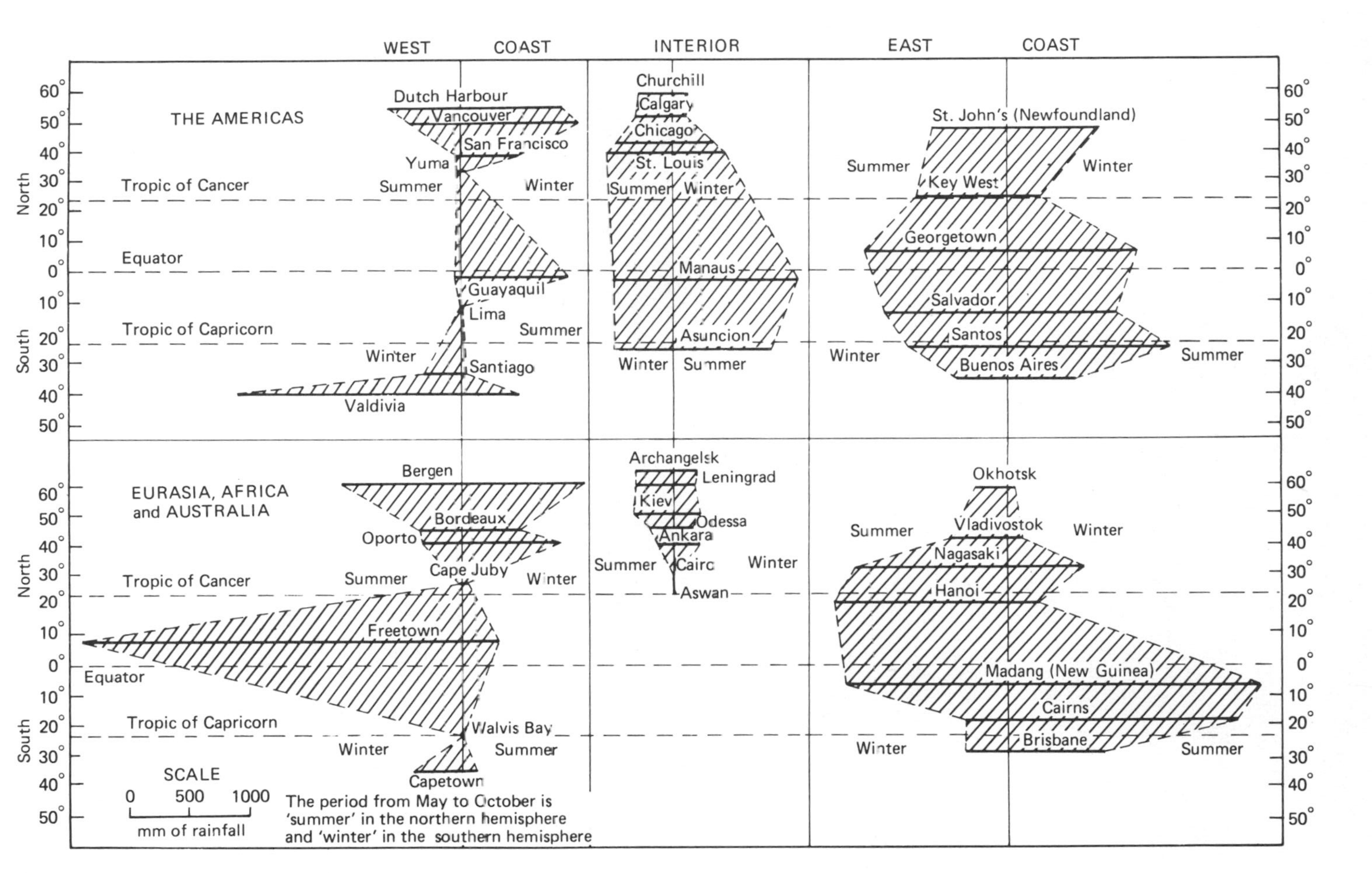

Fig. 2.22(a) Seasonal rainfall regimes

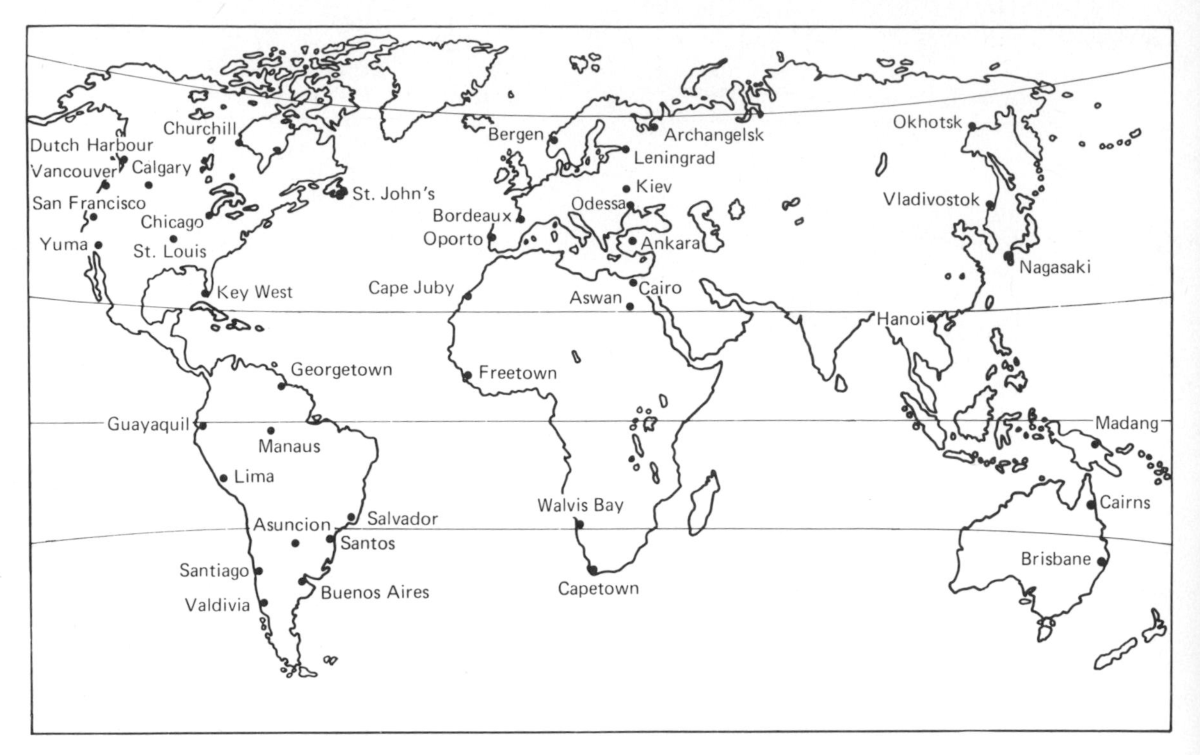

Fig. 2.22(b) Location map

in the continental interior, and on the east coast respectively. The lower half of the figure illustrates the seasonal rainfall regimes of similar locations in the remaining continents.

In the north of the west coast of North America both Dutch Harbour and Vancouver have high rainfall totals with a winter maximum, reflecting their location in the track of the westerlies and depressions from the Pacific. At San Francisco, however, rainfall is almost confined to the winter season, and, further south still, at Yuma, there is very little rain at all, this area being dominated by the stable air of the subtropical high pressure belt. With a cool ocean current offshore (Fig. 2.7), maritime tropical air from the sea is made stable by cooling from below. Rainfall increases markedly near the equator at Guayaquil. The sequence in the southern hemisphere from south to north is very similar. Valdivia, with its heavy winter rainfall, corresponds to Dutch Harbour and Vancouver, and Lima corresponds to Yuma.

The interior and the east coast of the Americas have generally higher rainfall totals than the west coast. In the northern interior, however, Churchill and Calgary have less rain than Dutch Harbour and Vancouver, and it occurs mainly in summer rather than winter. Here, the western cordillera hinders the entry of climatic influences from the Pacific and winters are dominated by the continental polar air mass of northern Canada. To the south, however, the northward drift of maritime tropical air from the Gulf of Mexico gives Chicago and St. Louis more rain than San Francisco and Yuma. Manaus has a higher rainfall total than Guayaquil on the west coast and its rainfall is more evenly distributed through the year. Rainfall totals are particularly high along most of the east coast of South America, from Georgetown to Buenos Aires. This is because of the maritime tropical air brought to this coast by the trade winds (Fig. 2.15). This air can penetrate inland to Asuncion but it cannot cross the Andes range to reach Santiago and Lima on the west coast. Valdivia owes its very heavy rainfall to the westerlies and depressions.

The situation in the other continents shows some similarity to that in the Americas. On the west coast of Europe and Africa, for example, Bergen is similar to Dutch Harbour. At about latitude 40°N Bordeaux and Oporto have more rain than San Francisco but, like San Francisco, they receive most of it in winter. Cape Juby (Morocco) corresponds neatly with Yuma. The heaviest rainfall on these coasts (e.g. at Freetown) is strongly concentrated in the northern summer. This is the result of the northward migration of the ITCZ and the consequent monsoonal inflow from the south Atlantic (page 113). South of the equator, Walvis Bay, like all west coast stations near the Tropics of Cancer and Capricorn, has very little rain. Further south, rainfall totals increase again, particularly in winter. In this respect Capetown corresponds to Santiago in South America. Africa does not extend far enough to the south to have a location with as wet a climate as Valdivia. The interior stations selected in Eurasia and Africa are located in eastern Europe, Turkey and north-east Africa. Archangelsk, Leningrad and Kiev have similar rainfall regimes to Churchill and Calgary but, south of here, Odessa and Ankara have much less rain than Chicago and St. Louis. They are much less affected by air of maritime origin than their North American counterparts. Cairo and Aswan, very arid locations in a continental interior, have no real counterparts in North America where the driest climates are found much nearer to the west coast. Cairo and Aswan are located near a source area for continental tropical air (page 113).

The east coasts of Asia and Australia resemble to some extent the west coasts of Europe and Africa in that very heavy rainfall occurs near the equator (Madang and Freetown). In eastern Asia and Australia, however, there is no equivalent of the desert climates at Cape Juby and Walvis Bay. In these latitudes, Nagasaki, Hanoi and Cairns have high rainfall totals. Also, in eastern Asia, rainfall totals decrease steadily northwards of latitude 30°N (Nagasaki, Vladivostok and Okhotsk) whereas, in western Europe, they generally increase northwards. Eastern Australia has a greater similarity with the east coast of South America, Brisbane and Cairns having similar rainfall totals and seasonal distributions to Buenos Aires, Santos and Salvador. Northwards in eastern Asia, however, rainfall totals decrease more rapidly, and the summer maximum tends to be more pronounced than in eastern North America. In eastern Asia, except at Nagasaki, very little rain falls in the winter season. This is the period of the winter monsoon when cold continental polar air flows outwards from the interior of Asia (Fig. 2.15(*a*)). In the summer season the flow of air is reversed and warm, moist air moves from the Pacific north-westwards towards northern Asia (Fig. 2.15(*b*)). Summer rainfall in eastern Asia and eastern Australia is also increased by the occurrence of typhoons (Fig. 2.20).

2.5 The classification of climates

A climatic classification is an attempt to arrange information concerning climate in a simplified and orderly form. Climatic elements such as temperature and precipitation in the different seasons of the year show great variation over the world. A major problem in evolving a classification is that climatic elements usually vary continuously over space, so that there are few sudden breaks in the sequence which can be used as boundaries between climatic types. Only where there is a sudden change in the relief of the land surface is there a sudden change in climate. Another problem is that, in the range of temperatures experienced at the earth's surface, the only very obvious significant value is that of the freezing point of water. For a climatic classification to be devised it is necessary to choose and justify other temperatures that can serve as rational boundaries between climatic types. Similarly, there is no objective definition of the characteristics of 'wet' and 'dry' climates in terms of annual or seasonal precipitation. In the development of systems of climatic classification, therefore, various techniques have been used to determine suitable values of temperature and precipitation for use as boundaries between climatic types. In some cases climatic types have been related to the distribution of vegetation and 'tropical forest', 'tropical savanna' and 'steppe' climates have been identified. Alternatively systems of climatic classification can be related to the factors that are responsible for the occurrence of the particular types of climate and climates have been classified in relation to influential air masses or pressure and wind systems.

THE KÖPPEN SYSTEM OF CLIMATIC CLASSIFICATION

This system uses temperature characteristics to

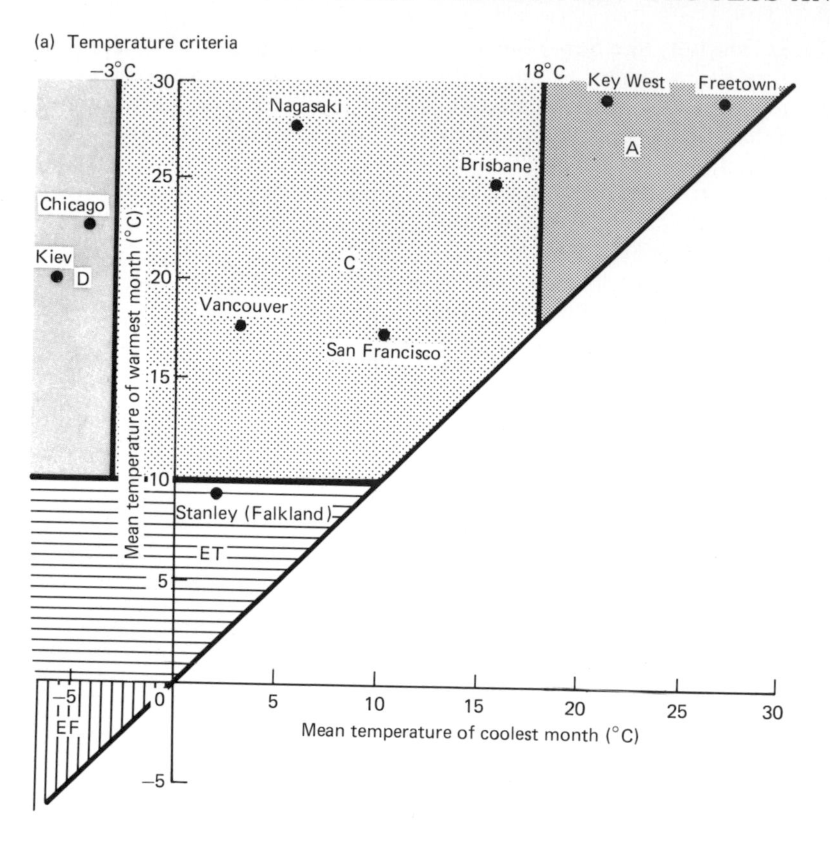

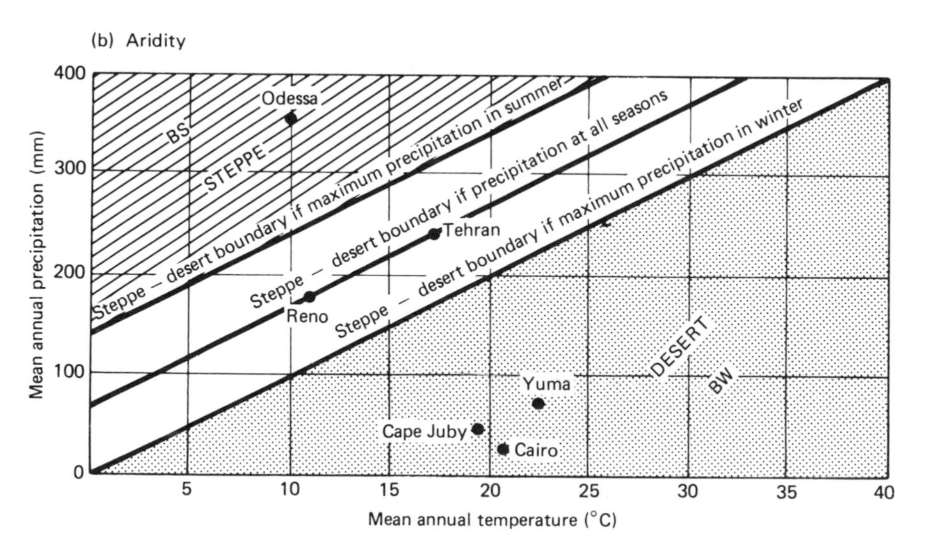

Fig. 2.23(a)(b) The Köppen system of climatic classification

identify four major climatic types in terms of temperature. These are termed A (tropical), C (mesothermal), D (microthermal) and E (tundra and ice cap) (Table I and Fig. 2.23(*a*)). The temperature characteristics of these four types are illustrated in Figure 2.23(*a*) and some representative locations, selected from those shown on the location map of Figure 2.22, are illustrated. In the Köppen system an A (tropical) climate has no mean monthly temperature below 18°C, so there is no real winter season. It also has a considerable annual rainfall. In a C (mesothermal) climate the warmest month has a mean temperature of at least 10°C and the coolest month lies between 18°C and −3°C. These are very wide limits and it is clear that Köppen's C climates include a very large variety of different types. In a C climate there are distinct winter and summer seasons as, for example, at Nagasaki (Fig. 2.23 (*a*)), where the warmest month reaches 27°C and the coolest month falls to 6°C. San Francisco, however, is more equable, with a warmest month of 17°C and a coolest month of 10°C. In Figure 2.23(*a*) the locations with the most equable temperature regimes are those that are located nearest the diagonal line, where there is no difference between the mean temperatures of the warmest and coolest months. A D (microthermal) climate has the same minimum temperature as a C climate for its warmest month (10°C) but its coolest month can be very much colder. The D climate also tends to have a greater difference in temperature between the warmest and coolest months. Chicago, for example, has a warmest month mean temperature of 23°C, which is almost as high as Brisbane's, but its coolest month falls to −4°C which is even colder than any month at Stanley in the Falkland Islands. E climates are those in which the mean temperature of the warmest month only reaches 10°C (in the ET type) and 0°C (in the EF type). The temperature boundaries used in the Köppen system can be justified as follows. The midsummer isotherm for 10°C is related to the poleward limit of tree growth. A mean monthly temperature of 18°C can be regarded as the minimum for a variety of tropical plants. The CD boundary (minimum mean monthly temperature of −3°C) is related to the length of the period of winter snow cover.

The B climates refer to arid and semi-arid areas where no permanent streams can develop. In this system of classification dry climates are not identified simply by reference to the amount of precipitation that occurs. Reference is also made to temperature which is related to the amount of evaporation that takes place (Table I). As indicated in Table I the inclusion of the letter 'h' indicates a mean annual temperature over 18°C and a 'k' indicates a mean annual temperature below 18°C. The maximum precipitation for a B climate also varies according to the season at which the precipitation occurs. If precipitation occurs mainly in summer a B climate can have a higher precipitation than if it occurs in winter, because evaporation is greater in summer than in winter. Figure 2.23(*b*) illustrates this same principle when it is applied to the two divisions of the B climate, BS (steppe) and BW (desert). The desert area expands into areas with higher precipitation totals if the maximum precipitation occurs in summer and it contracts into drier areas if most precipitation occurs in winter.

Supplementary letters are also added in the case of A, C, D and E climates. One group of these (f, s, w and m) refers to the season of precipitation. The other group (a, b, c and d) gives some indication of monthly temperature variations in C and D climates (Table I). In Figure 2.23(*a*) it can be seen that no maximum temperature for the warmest month is shown in the case of C and D climates, and no minimum temperature for the coolest month is shown for D climates.

Figure 2.23(*c*) shows the climatic regions of Australia according to the Köppen system. This map contains a wealth of information concerning Australia's climate. Desert areas (BW) occur over a very large area in central Australia and these are almost surrounded by rather less arid steppe areas (BS). In all these drier areas the mean annual temperature is over 18°C (h). Along the north coast and the northern part of the east coast mean monthly temperatures never fall below 18°C (A) and usually there is a dry season in winter (w). Near Cairns, however, this dry season is very short (m) and there is precipitation for most of the year. Inland from Cairns and southwards along the east coast winter temperatures fall below 18°C (C) but near Rockhampton and Brisbane, and extending along inland areas to the south, mean temperatures for the warmest month are over 22°C (a) but they fall below 22°C in the Melbourne area and Tasmania (b). Along the east coast from Rockhampton to Melbourne and Tasmania there is all-season precipitation with no dry season (f). South-west Australia, near Perth, and the

Table I Summary of the Köppen system of climatic classification

DRY CLIMATES	OTHER CLIMATES
Precipitation characteristics (first two letters) BS steppe climates BW desert climates *Temperature characteristics* (third letter) h mean annual temperature over 18°C k mean annual temperature under 18°C	*Temperature characteristics* (Fig. 2.23 (a)) (first letter) A tropical C mesothermal (warm temperate) D microthermal (cold) E tundra (ET), ice cap (EF) *Precipitation characteristics* (second letter) f no dry season (A, C and D climates only) s summer dry season w winter dry season m short dry season (A climates only)
Examples (see Fig. 2.22) Af Georgetown, Madang, Santos, Salvador Aw Key West, Guayaquil, Am Freetown, Manaus, Cairns	*Details of temperature* (C and D climates only) (third letter) a warmest month over 22°C b warmest month below 22°C c less than 4 months over 10°C d coldest month below −38°C
BWh Yuma, Lima, Cape Juby, Cairo, Aswan BWk Walvis Bay BSk Odessa, Ankara	
Cwa Hanoi, Asuncion Cfa Nagasaki, Buenos Aires, Brisbane Cfb Vancouver, Bordeaux, Bergen, Valdivia Cfc Dutch Harbour Csb San Francisco, Oporto, Santiago, Capetown	
Dfa Chicago, St. Louis Dfb Calgary, St. John's, Leningrad, Kiev Dwb Vladivostok Dfc Churchill, Archangelsk Dwc Okhotsk	

Adelaide area have temperatures similar to those of south-east Australia (C) but here there is a dry season in summer (s).

THE THORNTHWAITE SYSTEM OF CLIMATIC CLASSIFICATION

As explained above, the Köppen system of climatic classification is to some extent related to the growth of plants. The Thornthwaite system also

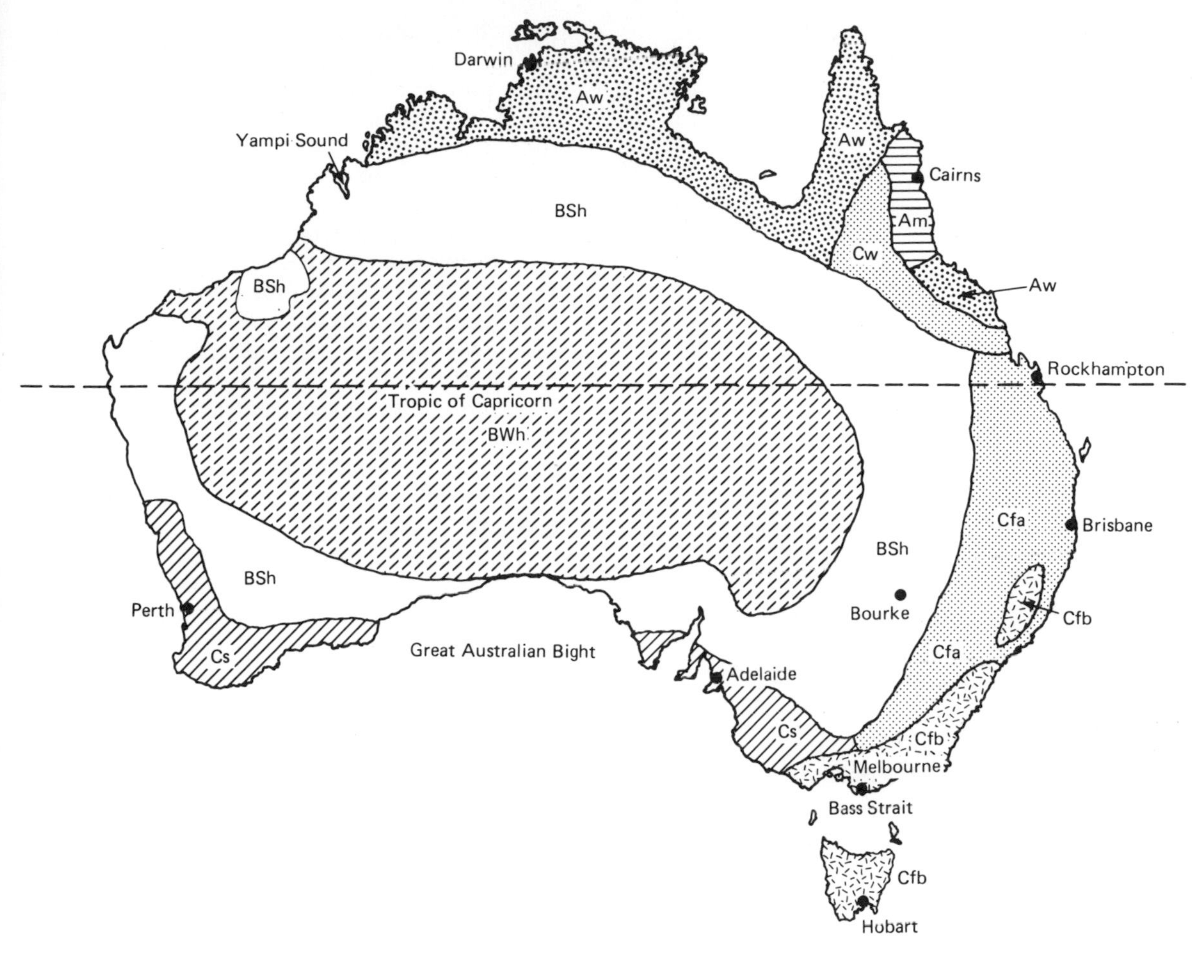

Fig. 2.23(c) Climatic regions of Australia (Köppen)

has a close relationship with plant growth since it is based upon the concept of potential evapotranspiration and therefore relates to the hydrological cycle (page 17, Figs 1.14(*b*) and (*c*)). Evapotranspiration consists of the evaporation of moisture from the soil and also transpiration by plants. Potential evapotranspiration, a basic concept in the Thornthwaite system, is the maximum amount of water that could be evaporated from the soil and transpired by plants if it were available. The general principle of the Thornthwaite system is that, in moist climates, precipitation over the year is greater than potential evapotranspiration, and therefore such climates have a moisture surplus. In dry climates precipitation is less than potential evapotranspiration. Dry climates are said to have a moisture deficiency. The potential evapotranspiration for each month of the year can be calculated from the mean monthly temperatures.

Figure 2.24 is a simple illustration of the application of the concept of potential evapotranspiration. Figure 2.24(*a*) represents a location where the maximum precipitation is in summer and Figure 2.24(*b*) represents one with the maximum precipitation in winter. In Figure 2.24(*a*) precipitation is greater than potential evapotranspiration (PE) from November to April and during this period surface run-off can occur. Precipitation is less than PE from May to October.

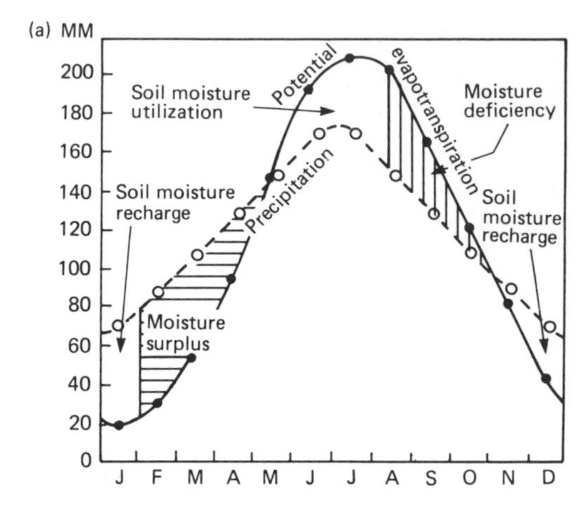

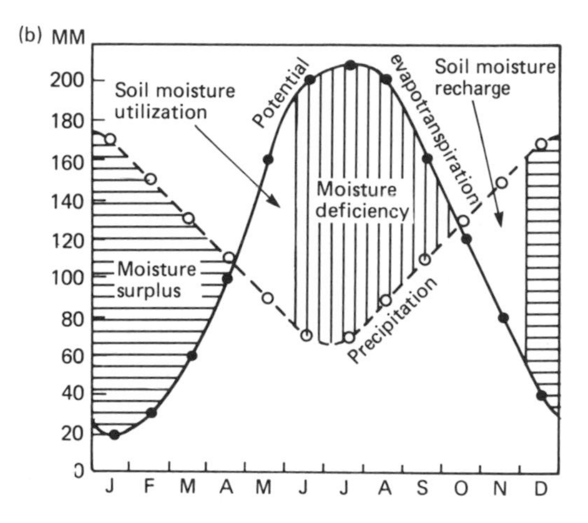

Fig. 2.24 Contrasting relationships between precipitation and potential evapotranspiration

In May PE begins to exceed precipitation because temperatures rise faster than precipitation. There is therefore a moisture deficiency. Until July this difference can be made up by the evapotranspiration of water that has been stored in the soil (soil moisture utilization) but from August onwards there is an absolute moisture deficiency which lasts until November, when precipitation, once again, exceeds PE. From November to January this moisture surplus is regarded as replacing the moisture that was removed from the soil in early summer (soil moisture recharge). By February there is once again a moisture surplus. This imaginary example illustrates the fact that, when potential evapotranspiration is considered, the season of heaviest precipitation need not be the time when most moisture is available. Figure 2.24(*b*) represents a location that has exactly the same potential evapotranspiration as Figure 2.24(*a*) and also has exactly the same total precipitation. In Figure 2.24(*b*), however, precipitation is concentrated in the winter season. In this case the moisture surplus in winter and the deficiency in summer are much greater. Despite the similarity of their precipitation and PE values conditions in the two locations are quite different.

The main elements of the Thornthwaite system are shown in Table II. The classification takes account of the moisture regime ((*a*) and (*c*) in the table) and also the thermal efficiency, measured by potential evapotranspiration ((*b*) and (*d*) in the table). Moisture conditions for the year as a whole are described by the Moisture Index (Table II (*a*)), A, B and C_2 being relatively moist climates and C_1, D and E being relatively dry. Then, in Table II (*c*), reference is made to seasonal variations in both moist and dry climates. The other major part of the classification (Table II (*b*) and (*d*)) refers to temperature (thermal efficiency) conditions as defined by potential evapotranspiration on both an annual, (*b*), and a seasonal, (*d*), basis. In (*b*) and (*d*) the implication is that the coldest climates (e.g. tundra and microthermal) have the greatest concentration of thermal efficiency in the summer season even though their annual thermal efficiency is quite small. In hot, megathermal climates, on the other hand, thermal efficiency is spread more evenly through the year.

Figure 2.25 shows some of the climatic variations in the USA as identified by the Thornthwaite classification system. Considering first the moisture regime, and taking the first letter of each group on the map, it can be seen that New York (B_3) has the most humid climate of the cities shown, and is followed by Seattle (B_2) and then Chicago and Miami (both B_1). Subhumid climates occur at Dallas (C_2) and also at Bismarck and San Francisco (both C_1) where the Moisture Index becomes negative. The driest climates are in the south-west at Los Angeles (semi-arid (D)) and Yuma and El Paso (both arid (E)). These arid and semi-arid climates at Los Angeles, Yuma and El Paso have little or no moisture surplus at any time of the year (d). This is also the case at Bismarck (C_1) which receives more precipitation in summer, when PE is greatest, than in winter. A somewhat similar case is illustrated in Figure 2.24 (*a*). On the west coast, Seattle (s) and San Francisco (s_2) have

Table II Summary of the Thornthwaite system of climatic classification

MOISTURE REGIME

(*a*) *Annual*

		Moisture Index
A	Perhumid	100+
B_4	Humid	80–100
B_3	Humid	60–80
B_2	Humid	40–60
B_1		20–40
C_2	Moist subhumid	0–20
C_1	Dry subhumid	0– –20
D	Semi-arid	–20– –40
E	Arid	–40– –60

$$\text{Moisture index} = \frac{S - 0.6D}{PE}$$

where S is the surplus in the wet season and D is the deficit in the dry season.

(*c*) *Seasonal*
For moist climates (A to C_2):

r	little or no moisture deficiency
s	moderate moisture deficiency in summer
s_2	large moisture deficiency in summer
w	moderate moisture deficiency in winter
w_2	large moisture deficiency in winter

For dry climates (C_1 to E):

d	little or no moisture surplus
s	moderate winter moisture surplus
w	moderate summer moisture surplus
s_2	large winter moisture surplus
w_2	large summer moisture surplus

Examples (see Fig. 2.25 overleaf)

New York	$B_3B'_1$ r b'_2
Seattle	$B_2B'_1$ s a′
Chicago	B_1 B'_1 r b'_2
Miami	B_1A' r a′
Dallas	$C_2B'_4$ s b'_4
San Francisco	$C_1B'_1$ s_2 a′
Bismarck	$C_1B'_1$ d b'_1
Los Angeles	D B'_2 d a′
Yuma	E A′ d a′
El Paso	E B'_3 d b'_3

THERMAL EFFICIENCY (POTENTIAL EVAPOTRANSPIRATION)

PE (mm) (*b*) *Annual thermal efficiency*

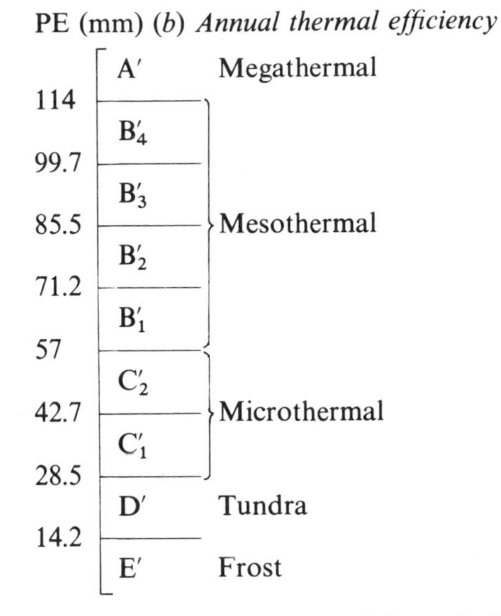

(*d*) *Summer concentration of thermal efficiency* (% of thermal efficiency in summer season)

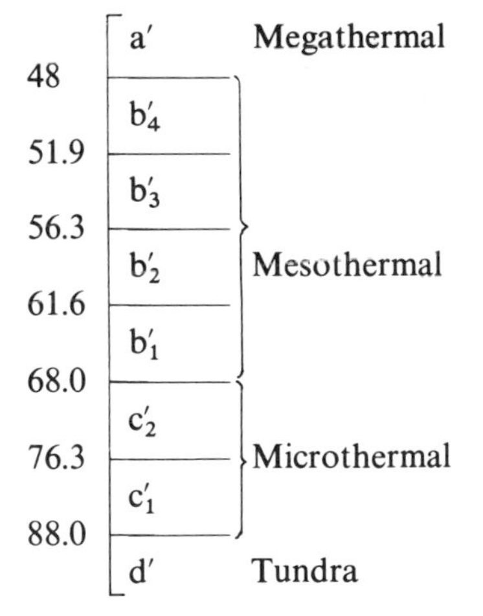

a moisture deficiency in summer, not because their summers are particularly hot but rather because a high proportion of their precipitation comes in winter. Dallas is an interesting case, with its moderate moisture deficiency in summer (s). This is more the result of its high summer temperatures (and hence its high PE) than a deficiency in precipitation at that season. The easterly locations, Chicago, New York and Miami, have little or no moisture deficiency at any season (r) despite their hot summers.

Yuma and Miami have megathermal climates (A′) in terms of thermal efficiency and there is little or no variation between the seasons (a′). The megathermal type of thermal efficiency distribution through the year (a′) also extends along the west coast to Los Angeles (B′), San Francisco (B'_1) and Seattle (B'_1). This indicates that these locations have unusually equable temperature regimes through the year. Normally, cooler climates such as these have a greater concentration of thermal efficiency in the summer (Table II (d)). El Paso (B'_3) and Dallas (B'_4) are cooler than Miami and Yuma and their summer concentration of thermal efficiency (Table II (*d*)) is appropriate to their potential evapotranspiration (Table II (*b*)). Bismarck, Chicago and New York all have thermal efficiencies at the lower end of the mesothermal range (B'_1) (Table II (*b*)). Of these, Bismarck has the greatest summer concentration of thermal efficiency (b'_1). In fact, Bismarck has the highest summer concentration of thermal efficiency of any of the places named in Figure 2.25, as befits a location near the centre of a great land mass.

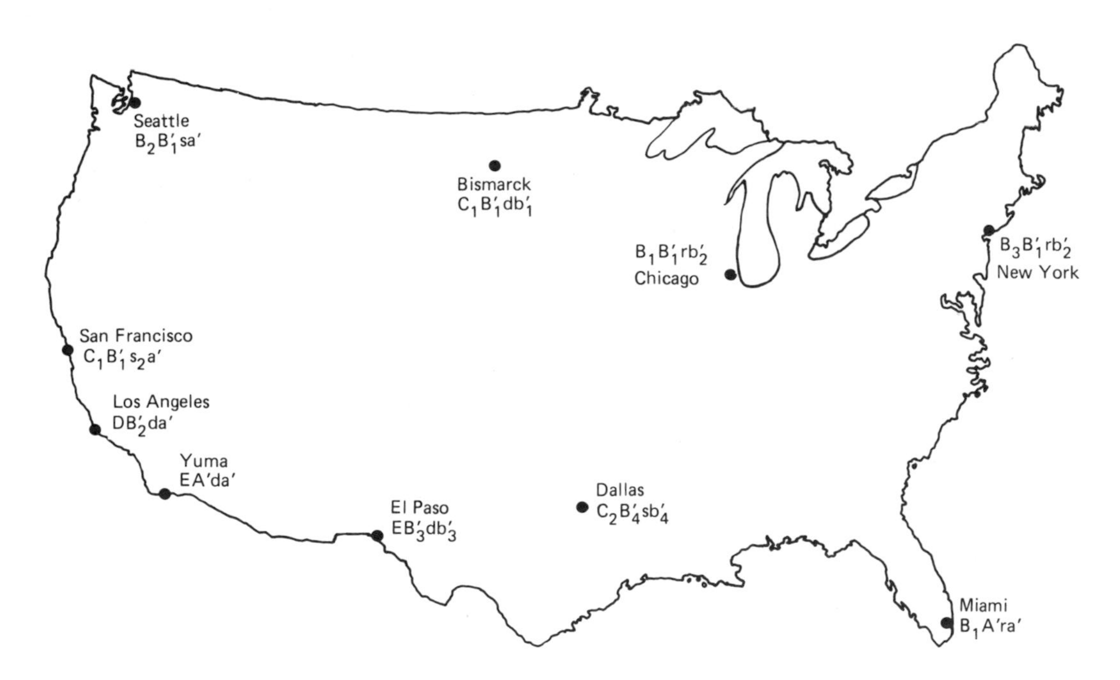

Fig. 2.25 Climatic variations in the USA (Thornthwaite system)

Exercises

1. Explain why
   (*a*) a cloudless sky by daylight is blue but at sunset it turns red;
   (*b*) the albedo of the earth's surface varies from place to place;
   (*c*) temperatures in the troposphere generally fall as one moves away from the earth and towards the sun;
   (*d*) temperatures tend to decrease towards the Poles;
   (*e*) the annual range of temperature tends to increase towards the Poles.
2. Discuss the influence of (*a*) winds and air masses, and (*b*) ocean currents, on summer and winter temperatures over the earth's surface.
3. Explain the ways in which the following can influence temperature:
   (*a*) relief features;
   (*b*) vegetation;
   (*c*) urbanized areas.
4. (*a*) What factors can cause a change in the relative humidity of the air?
   (*b*) Explain how meteorological conditions can influence the stability of the atmosphere.
5. Describe and explain
   (*a*) the relationships between incoming solar radiation and outgoing terrestrial radiation over the earth's surface;
   (*b*) the exchanges of water that take place between land areas, oceans and the atmosphere in the global hydrological system.
6. Describe and account for the distribution of the earth's major pressure and wind systems.
7. Give an explanatory description of the possible effects on the weather in the United Kingdom of the following air masses:
   (*a*) continental polar;
   (*b*) maritime polar;
   (*c*) maritime tropical.
8. (*a*) How do depressions influence the weather of the areas over which they pass?
   (*b*) Describe and explain the processes that operate in depressions.
9. 'Precipitation over the earth's surface tends to follow the sun.' Explain and discuss.
10. Figures 2.26 and 2.27 describe the weather at various recording stations in the Mojave Desert in California for a period in late December and early January.
    (*a*) Figure 2.26 shows the variations in temperature that occurred at two weather stations on a single day. On this day the sun rose slightly before 8 am and set shortly after 4 pm. At station A the sky was clear for the whole day. At station B however stratus cloud covered the sky from about 5.45 am onwards and, at about 11 am, a gale developed and caused clouds of dust.
    Describe and suggest reasons for the differences between the two thermograph traces.
    (*b*) Figure 2.27 shows the location of different weather stations in an area of considerable relief. Describe and suggest reasons for the differences in their mean maximum and mean minimum temperatures for the

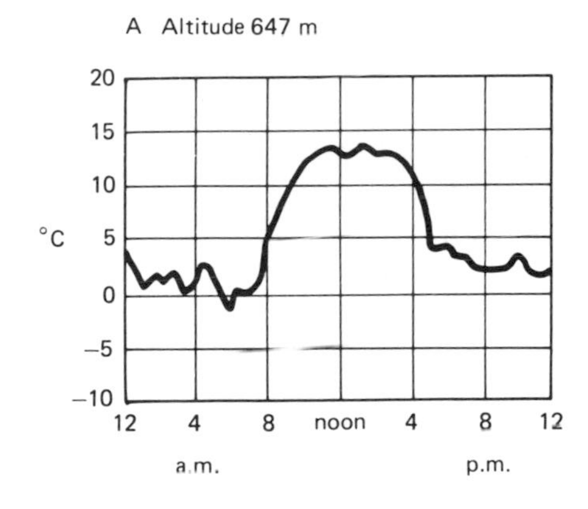

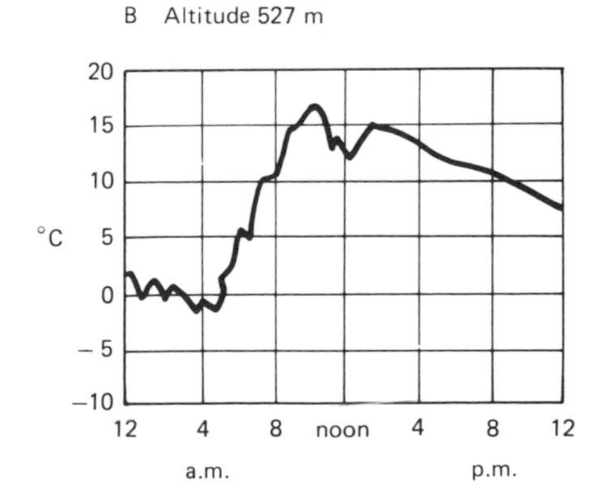

Fig. 2.26 Thermograph traces for stations in a desert basin

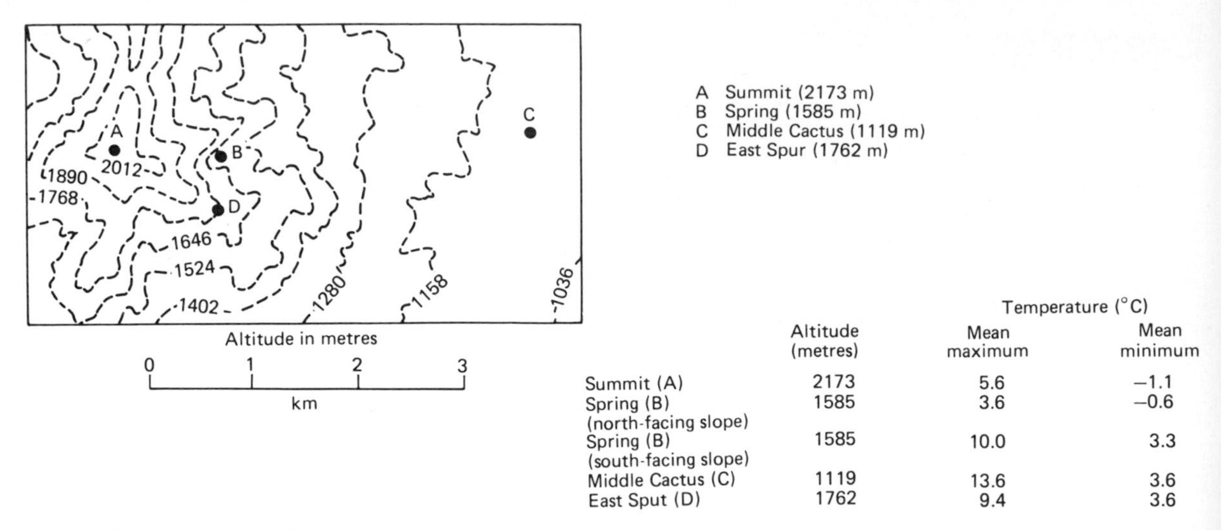

	Altitude (metres)	Temperature (°C) Mean maximum	Mean minimum
Summit (A)	2173	5.6	–1.1
Spring (B) (north-facing slope)	1585	3.6	–0.6
Spring (B) (south-facing slope)	1585	10.0	3.3
Middle Cactus (C)	1119	13.6	3.6
East Sput (D)	1762	9.4	3.6

Fig. 2.27 Location of four weather stations

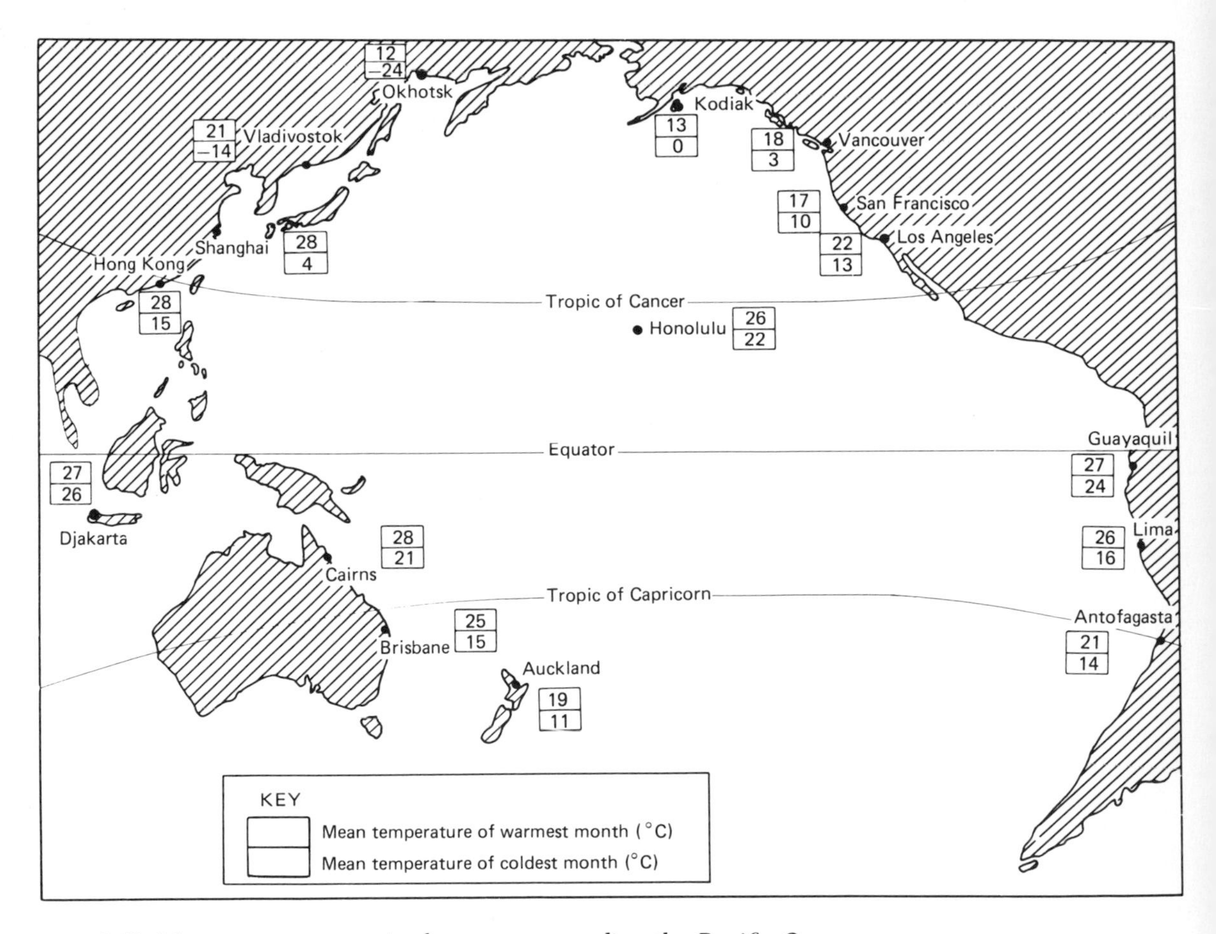

Fig. 2.28 Mean temperatures in the area surrounding the Pacific Ocean

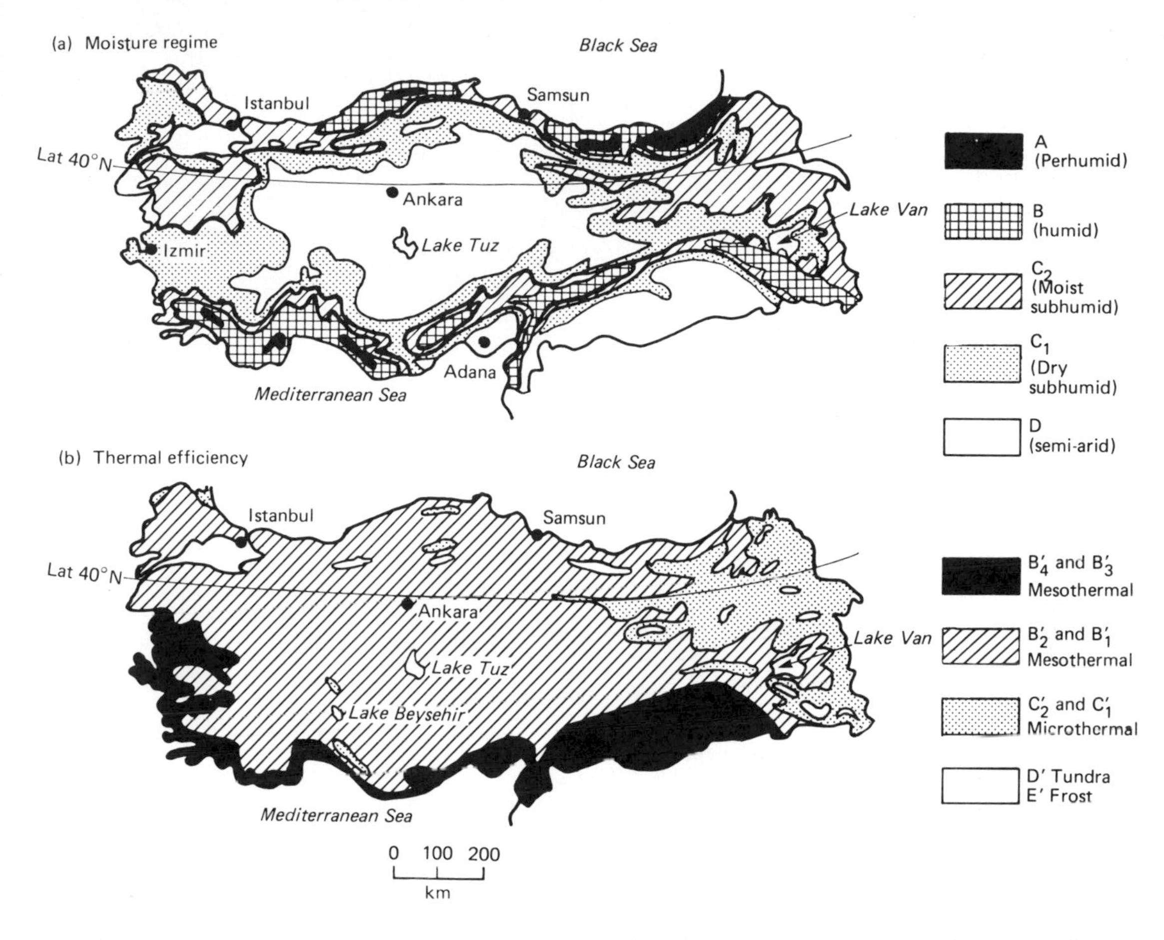

Fig. 2.29(a)(b) Spatial variations of temperature and moisture conditions in Turkey according to the Thornthwaite system

period during which observations were taken. Note that there are two sets of figures for Station B.

11. Describe and suggest explanations for the temperature variations that are illustrated in Figure 2.28.
12. The maps (Figures 2.29(*a*) to (*d*)) show the spatial variations of temperature and moisture conditions in Turkey, according to the Thornthwaite system of climatic classification.
    (*a*) Compare the characteristics of the climates of Samsun, Istanbul and Adana.
    (*b*) What deductions could you make concerning the influence of relief features on the climate of Turkey?
    (*c*) What problems and opportunities would you expect the climate of Turkey to present in relation to the development of agriculture? Illustrate your answer with a sketch map.

(c) Seasonal variation of effective moisture

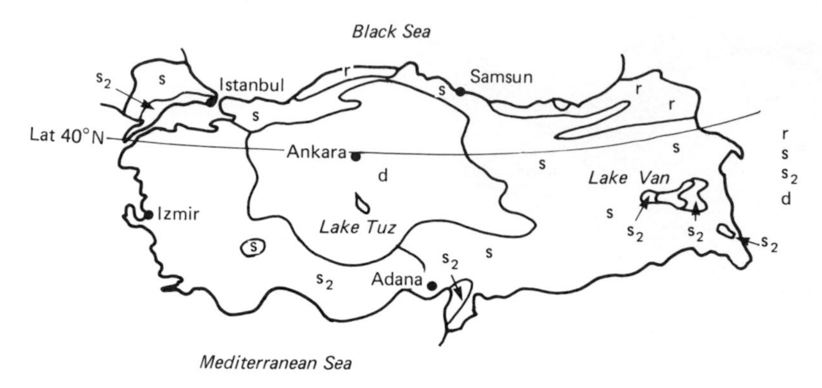

(d) Summer concentration of thermal efficiency

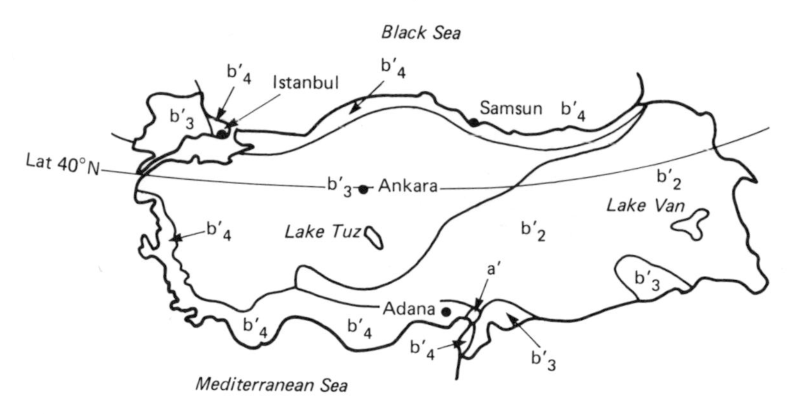

Fig. 2.29(c)(d)

13. Figure 2.30 illustrates the mean monthly temperature and the mean monthly precipitation at five different locations which are referred to as A, B, C, D and E.
    (*a*) Give an answer A, B, C, D or E to each of the following questions. Which location has:
      (i) the smallest annual range of temperature?
      (ii) the greatest annual range of temperature?
      (iii) the greatest variation in mean monthly precipitation?
      (iv) the smallest variation in mean monthly precipitation?
    (*b*) The five locations (A to E) whose climatic conditions are illustrated in Figure 2.30 are, not necessarily in the same order, Cape Town, Moscow, New York, Poona (India) and Scilly (England). State which letter (A to E) refers to each of these locations.
    (*c*) Explain the reasons for the differences between the climates of Cape Town, Moscow, New York, Poona and Scilly.

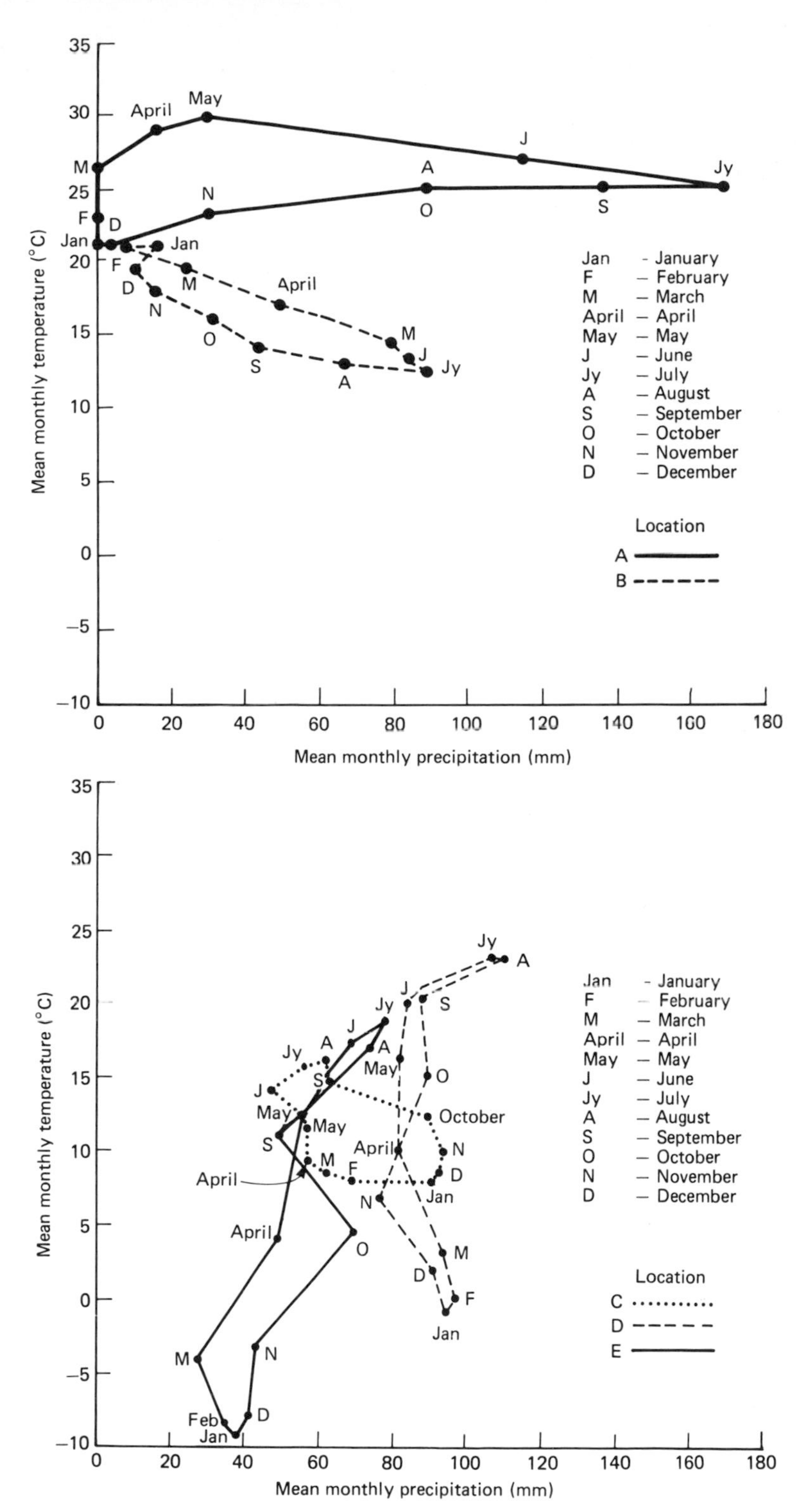

Fig. 2.30 The mean monthly temperature and the mean monthly precipitation at five different locations

3 Vegetation

The study of vegetation (plant geography) is one aspect of biogeography. Biogeography also includes the study of animal life but this has been given much less attention than plant geography. This may be because the distribution of different kinds of vegetation seems to relate more closely than the distribution of animals to patterns of relief and climate. Also, the characteristics of vegetation are as easy to observe in the field as relief features and weather. The study of soils forms another aspect of biogeography. Soils, too, display variations from place to place over the earth's surface which are related to environmental conditions. They form another branch of biogeography.

3.1 General principles

Ecosystems

An ecosystem is an ecological system in which interaction takes place between living things and their non-living environment. It is a system in the same sense as a slope system (Fig. 1.13(*e*)) or a hydrological system (Figs. 1.14(*b*) and (*c*)). There are inputs and outputs and various types of exchanges within the system. Climate has an influence on vegetation, and both climate and vegetation have an influence on the characteristics of the soil and also on the types of animals. Animals, in turn, can influence the vegetation and the soil, and the type of soil can influence the characteristics of the fauna within the soil. Ecosystems can be identified and studied at any scale from a local patch of woodland to the whole of the earth.

Several different elements make up an ecosystem. First of all there is its inorganic base, consisting of air, water, soil, and rock. Then there are the producers, algae and leafy plants and trees which produce food by photosynthesis, using the energy of light. This energy is transferred to herbivores that feed on plants and to carnivores which feed on animals. Dead plants and animals are consumed by decomposers. In this way they add organic material to the soil which can be used by trees and other leafy plants. In effect, the input of solar energy by photosynthesis is distributed through the system. In this process the ecosystem produces a considerable amount of organic material in the form of plants and animals. This is referred to as the biomass. In general, forest ecosystems produce the greatest biomass, particularly those located in hot, wet equatorial regions. This is because of the high temperatures and the abundance of water. The high levels of insolation supply a great deal of energy to all parts of the ecosystem. Deserts and areas of tundra produce much smaller quantities of biomass. Five major ecosystems, identified in terms of plant formations, are forest, savanna, grassland, desert and tundra.

Climatic climax

The term 'climatic climax', when used in relation to vegetation, means that the vegetation has attained a state of equilibrium in relation to conditions of soil and climate. Such a vegetation used to be termed 'natural vegetation'. This stable condition has been reached over a long period of time during which successive plant communities have occupied the area. The climax vegetation is made up of those plants that have competed most successfully. Detailed characteristics of the climatic climax vegetation may vary from place to place in response to variations in soil conditions. Such variations may be termed 'edaphic climaxes'. The climax vegetation may also vary according to local conditions of slope and drainage. The climax vegetation is not identical for all parts of the world that have a particular type of climate. Different areas will have different species of plants even though their climates are almost identical. This is because there are many barriers on the earth's surface to the diffusion of plant species, in particular the great oceans. Such areas, however, even though they have different plant species, tend to have vegetation with similar general physical characteristics (life forms). Tropical rain forests, for example, in Africa and South America, have similar life forms, but different plant species because of the barrier of the Atlantic Ocean. Many different factors tend to delay or to prevent the devélopment of a climatic climax vegetation. The climate of an area may change, so that the vegetation begins to readjust itself to the new environmental condition. Also, volcanic activity, processes of erosion and deposition, and the occurrence of fires, may destroy the evolving vegetation. The evolution of the vegetation may also be influenced by the activities of man in the exploitation of the resources of forests and the development of land for agriculture.

Seres

A sere is a sequence of events by which the vegetation of an area evolves with the passage of time. It refers to the invasion of an area by a succession of plant communities until eventually a stable condition, the climatic climax, may be attained. Several kinds of sere can be identified. A *prisere* comprises the complete chain of successive plant communities that develop in an area, from the early pioneers to the climatic climax vegetation. When the prisere is completed the ecosystem is closely related to the local climatic conditions. Each stage of a prisere is referred to as a seral community, and each successive seral community tends to create environmental conditions that are more favourable to, and thus permit colonization by, the succeeding, more advanced, seral community.

The prisere may begin on a newly created land surface such as a beach, a sand dune or a lava flow, on which there is no soil. Primitive plants which can tolerate dry conditions and extreme temperatures invade. Gradually, a rudimentary soil is created and the early pioneer plants are replaced by more advanced types. These may increase the humus content of the soil and also, possibly, provide shady conditions which make possible the development of seedlings of more advanced plants. Animals may enter the area and interact with the plants. Ultimately the climatic climax vegetation comes into existence, and the ecosystem is in harmony with the local climatic conditions. Many priseres in different parts of the world are in the process of developing towards the climatic climax but have not yet reached it. The volcanic island of Krakatau in the East Indies erupted in 1883 and its surface became covered with volcanic ash; its tropical forest vegetation being destroyed com-

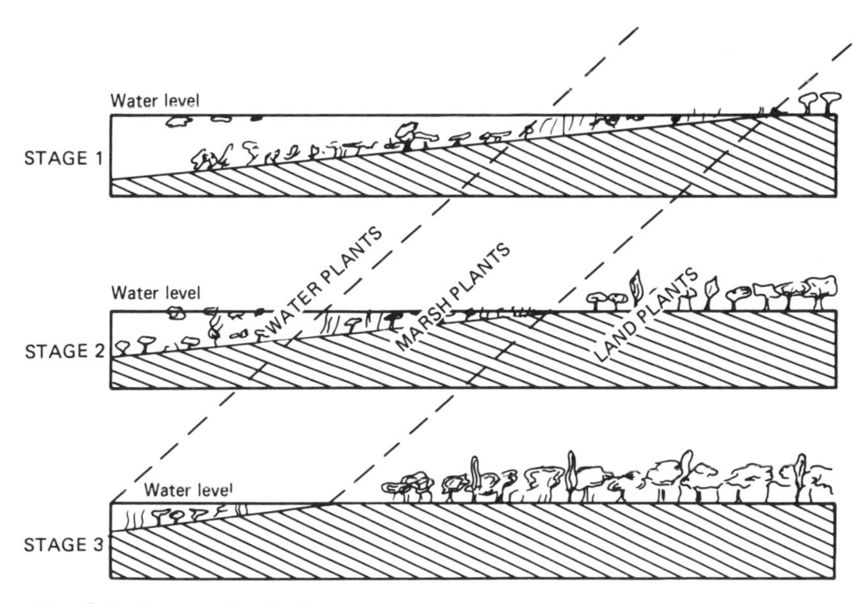

Fig. 3.1 Stages of a hydrosere

pletely. The tropical forest vegetation could not immediately re-establish itself, but, within twenty years, the new land surface had acquired a cover of savanna vegetation and this has since been succeeded by tropical rain forest. In a prisere each stage creates more favourable conditions for the establishment of more complex plant communities.

Priseres may develop in a variety of different environments. If development takes place on a very dry surface, such as rock or a sand dune, the prisere is referred to as a *xerosere*. On the rock surface the prisere can be termed specifically a *lithosere* and on the sand dune it is a *psammosere*. This latter type of prisere often involves the development of marram grass (page 85). Another special type of prisere is the *hydrosere* that develops in wet conditions in a shallow lake or a swamp. Early plant colonizers here are reeds. The various marsh and water plants trap sediment and eventually the lake or swamp becomes dry ground which can be invaded by a greater variety of plants, including trees (Fig. 3.1). A special kind of hydrosere is the development of a salt marsh on a coastline (page 84, Fig. 1.52(*g*), and page 86, Fig. 1.53). Because of the presence of salt water this is called a *halosere*. A *plagiosere* is said to exist where man has interfered with the development of the climatic climax vegetation, by, for example, clearing trees and developing an economy based on grazing, which prevents the regeneration of forest. This eventually may lead to a plagioclimax vegetation which is quite different from the climatic climax. On the other hand, if grazing ceases, a subsere may begin which may eventually lead to the creation of a climatic climax vegetation.

3.2 World vegetation types

GENERAL PRINCIPLES

The climatic climax vegetation or natural vegetation is the type of vegetation that would have developed if there had been no interference by man. In many parts of the world, because of man's activities in farming and the exploitation of forests, the natural (climatic climax) vegetation does not exist. In constructing a natural vegetation map of the world (Fig. 3.2) such areas are allocated the climax vegetation that they are assumed once to have possessed. There are five major vegetation types. These are forest, savanna, grassland, desert, and tundra. They are sometimes referred to as *biochores*. They are characterized by different types of plant 'life forms' such as trees, shrubs and grasses which thrive under different climatic conditions (Fig. 3.3) and which are limited in their spatial distribution by climatic conditions. These major biochores may be subdivided into various 'formation types'. Forests, for example, may be deciduous or evergreen, broad leaved or needle leaved. Finally, in greater detail still, formation types may be subdivided into 'plant communities' or 'associations' which have adapted themselves to certain detailed site conditions. Different plant communities are associated, for example, with limestone areas, sand dunes, and salt marshes.

At the world scale climate is the most important influence on vegetation type. Plants take up water by their roots, use it for photosynthesis and give it off through their leaf pores (stomata) by transpiration. Transpiration tends to be greater in conditions of high temperature and low humidity. Large thin leaves tend to lose more water by transpiration than waxy leaves or needles or spines. Hence, the water requirements of different plants vary. Forest vegetation generally occurs in areas where there is a considerable amount of precipitation, but much depends on the relationship between precipitation and evapotranspiration. The detailed type of forest depends partly on the seasonal distribution of precipitation. In dry climates the vegetation is described as 'xerophytic'. It has characteristics which permit it to survive where water is in short supply. Frequently, transpiration is reduced to a low level or the root system is particularly efficient in obtaining supplies of water. Temperature is also important. The great majority of plant species are suited to a hot climate. Species are particularly numerous near the equator and they decrease towards higher latitudes. In cold climates plants have defensive characteristics which allow them to survive.

At a smaller scale vegetation is greatly influenced by the detailed conditions of its site. Low ground in the bottoms of valleys, for example, may have a greater supply of soil water than quick-draining slopes close by because the water table is nearer to the ground surface (Figs. 1.15 and 1.16). Also, insolation may increase evaporation from

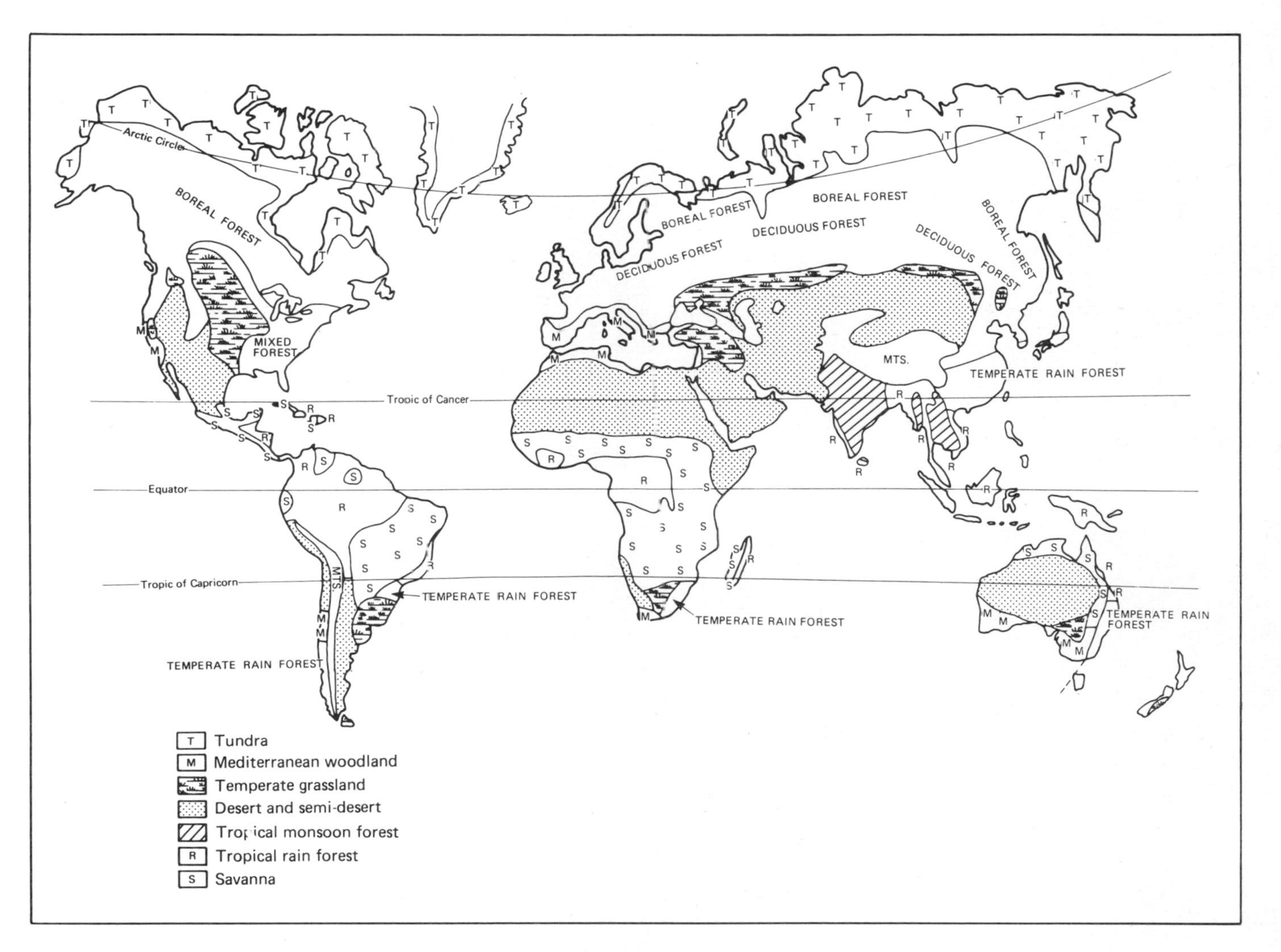

Fig. 3.2 Generalized world distribution of types of vegetation

valley slopes. South-facing slopes in the northern hemisphere tend to have drier soils than north-facing slopes.

Because of these relationships between vegetation and climate, plants are subdivided according to their climatic requirements. In respect of their moisture requirements they are divided into hygrophytes which require wet conditions, mesophytes, and xerophytes, that can survive in dry areas. This classification refers not only to general world climatic conditions but also to detailed site conditions. In respect of temperature they are divided into megatherms (adapted to high temperatures), mesotherms, and microtherms (adapted to cold conditions) (see Table I and II).

FORESTS

Forests have the highest productivity of all the biochores since most of them are located where temperatures are reasonably high and there is a surplus of precipitation over evapotranspiration.

Tropical rain forest

Tropical rain forests are found mainly in lowland areas near the equator such as the Amazon Basin in South America, the Congo Basin in Africa, and the islands of the East Indies (Fig. 3.2). In some areas, particularly central Africa, Ceylon and some East Indian islands, much forest has been cleared for agriculture. In south-east Asia and eastern Australia, in particular, forest extends beyond the tropics. All these areas have a heavy, reliable rainfall, with possibly a short drier season which is insufficient to reduce supplies of soil water significantly (Fig. 3.3(*a*)). Hence, growth is continuous through the year. As a result, nearly all the trees are broad-leaved evergreens, but the detailed species in the three major areas are different because the vegetation in each area evolved separately. Since there are no clear-cut seasons, germination, growth, flowering and fruiting can all take place at any time of the year. The forest is therefore evergreen. Within the forest there is a considerable range of microclimates, with variations in temperature, exposure to sunlight, humidity and wind speed. The main morphological feature of the tropical rain forest is the stratification of the vegetation. Leaves tend to be clustered at various heights above the forest floor. The tallest trees (emergents), up to 40 m high, project above the general level. Beneath these are a second layer, at about 15 to 30 m, and a third layer at about 10 m. The tallest trees commonly have umbrella-shaped crowns but the smaller ones tend to be conical in shape, with rather narrow crowns. Nearly all the trees have straight trunks with relatively few branches below the crown. Young seedlings of these major trees can be found on the forest floor. These grow rapidly if a major tree dies. Because of the high level of humidity in the lower layers of the forest, trees generally have thin barks. The tree needs no protection from the weather. Pointed drip-tip leaves to shed excess moisture are also common. Flowers and fruit even form on tree trunks, a characteristic termed 'cauliflory'. Some major trees have buttress shaped roots extending up the lower trunk for a distance of up to 8 m (plank buttresses). These tend to give the tree added stability. Lianas wind from tree to tree from the forest floor, beginning their growth in shaded conditions and forming branches when they reach the upper layers. *Epiphytes*, such as ferns and orchids, grow in crevices in the bark of the larger trees. Parasites grow in similar positions but also are nourished by the host tree. Some epiphytes have aerial roots in which falling leaves may be trapped and thus provide the plants with nutrients. In some cases stranglers begin their growth on higher branches and send their roots down to the ground, eventually enclosing the host tree. On the densely shaded forest floor there are relatively few plants, but there is much decaying plant debris on which saprophytes feed.

On river banks and other open areas there is a dense growth of vegetation, mainly because in these locations sunlight can reach a low level. Frequently a hydrosere (page 140) has developed on the margins of rivers and lakes, consisting of a succession of reeds, mangroves and palms. In areas where the forest has been cleared by man for agriculture or other purposes, and then abandoned, a subsere (page 140) develops. Low growing shrubs invade the area and eventually climatic climax rain forest may be re-established.

Tropical monsoon forest

This type of forest occurs mainly in India, Burma, Thailand and Kampuchea, but a very similar type of vegetation is found in northern Australia and parts of west Africa.

Tropical monsoon forest appears to have some resemblance to tropical rain forest but the major trees are not so high, there are fewer epiphytes, and

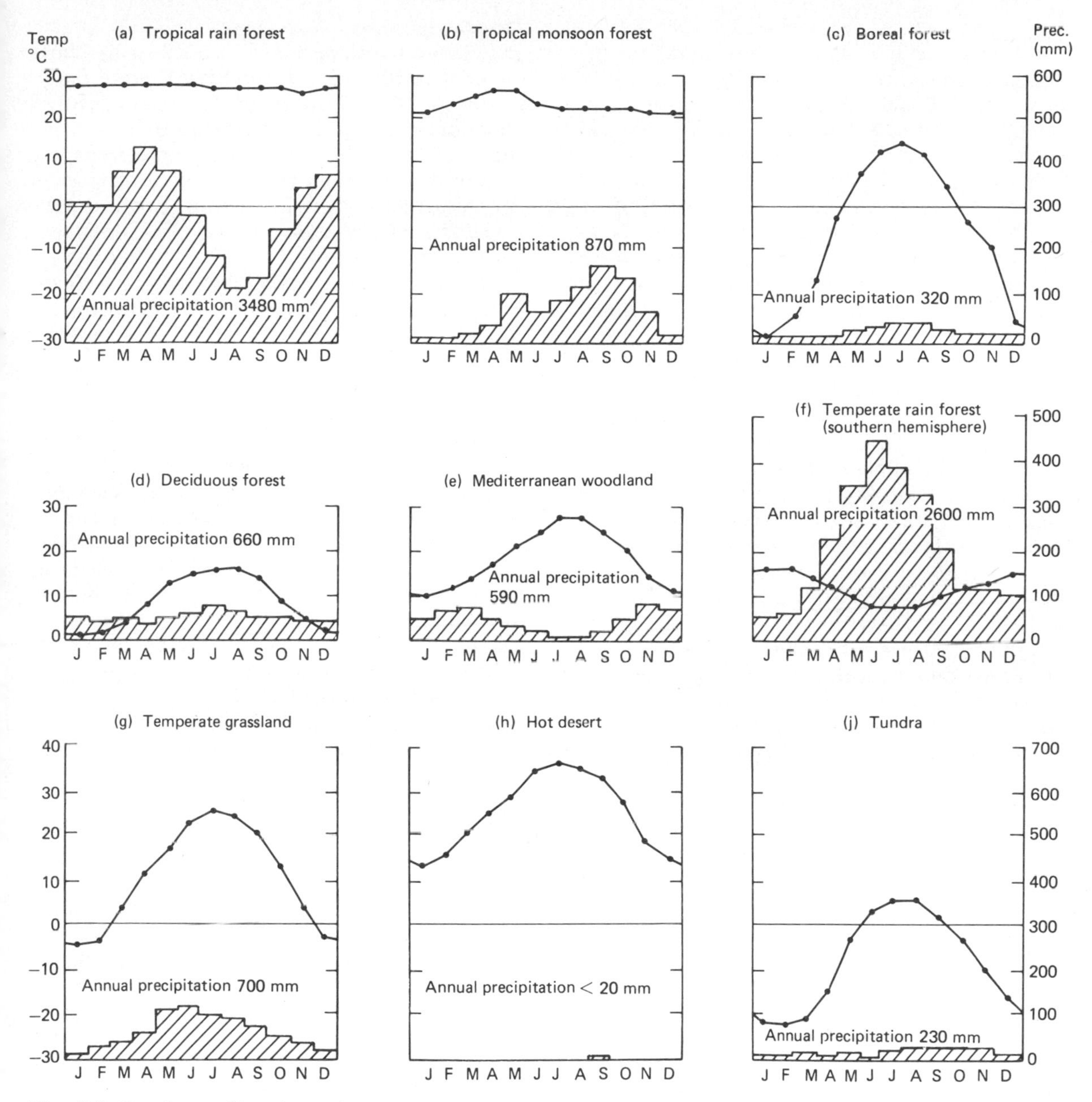

Fig. 3.3 Specimen climatic regimes

branching takes place at much lower levels. The upper canopy consists of the rounded crowns of the major trees. These trees, in contrast to those of the tropical rain forest, have a thick, rough bark. Usually there is also a fairly dense lower storey which often includes bamboo. The climate here is marked by a distinct dry season (Fig. 3.3(*b*)) and most trees are deciduous.

The areas in which tropical monsoon forest occurs have a long history of human occupation. Relatively advanced civilizations developed very early in south-east Asia. It is therefore possible that this vegetation type is a plagioclimax (page 140) which has been modified by human activity. Teak, one of the commonest trees, is resistant to fire.

Boreal forest

The boreal (coniferous) forest is confined to the northern hemisphere. In the southern hemisphere there are few land areas in these latitudes. It extends in a broad belt across North America and Eurasia (Fig. 3.2).

The climate is characterized by very long cold winters and short cool summers (Fig. 3.3(*c*)). In only four or five months does the mean temperature rise above 6°C, the normal minimum for plant growth. Although, in places, annual precipitation totals are very low, most rain comes in summer (Figs. 2.21(*b*) and (*c*)) when the higher temperatures allow it to be used by plants. Summers are generally so cool that there is little loss of soil moisture through evaporation. In winter soil moisture is mostly frozen. This means that there is a marked annual climatic rhythm.

There are relatively few species of trees, and many stands are of a single species. Most of the trees are evergreen and hold their leaves through the winter, thus allowing them to start new growth as soon as temperatures rise in spring, thus making the fullest possible use of the short growing season. Needle-shaped leaves with thick cuticles (outer leaf coverings) and with a very small surface area, protect the trees from loss of moisture by transpiration when the soil is frozen in winter. Even in summer the relatively low temperatures mean that there is little evapotranspiration. To some extent the spatial distribution of tree species is related to soil characteristics. Pines tend to be more numerous on thin, sandy soils, and spruce are more numerous on clay soils with a greater moisture content. Deciduous trees such as birch and larch tend to be common in the coldest areas, such as northern Siberia. Most of the trees have straight trunks and a conical shape, which tends to shed snow rapidly (thus avoiding the breakage of branches) and also tends to give stability in high winds. Trees cast a heavy shade so that few plants are able to grow at or near ground level. Usually there is only a layer of moss. Tree roots often penetrate the soil only to a shallow depth because of the presence of permafrost (permanently frozen soil or subsoil). If an area of boreal forest is cleared and then abandoned, a subsere begins. The area tends first of all to be invaded by deciduous birch, then later by evergreen conifers which grow to a greater size and cause the deciduous species to be 'shaded out'.

Deciduous forest

Deciduous forest occurs in Eurasia and North America to the south of the boreal forest. On its drier side (south in Eurasia, west in North America) it is succeeded by temperate grassland. Many of these areas, particularly in Europe and eastern North America, have been developed for agriculture and industry and have become densely populated. Hence, little of the original deciduous forest is left.

Precipitation here is generally greater than in the boreal forest and it is fairly evenly distributed through the year (Fig. 3.3(*d*)). Mean monthly temperatures fall below 6°C in winter, thus giving a check to vegetation growth, but this cool season is shorter than in the boreal forest and minimum temperatures are considerably higher (Figs. 3.3(*c*) and (*d*)). In this winter cool season tree roots are unable to take up moisture, so the trees shed their leaves so as to reduce the loss of moisture by transpiration. Leaf buds remain dormant through the winter and then break into growth when the temperature rises in spring. At this season, low-growing herbaceous plants and shrubs come into leaf and flower before the trees develop leaves and cast shade on the forest floor. The luxuriance of this lower layer depends to some extent upon the type of tree in the forest and the amount of shade that it casts. Varieties of trees depend to some extent upon soil conditions. In Europe various types of oak are common. Ash and beech tend to favour well-drained calcareous soils. Birch tends to occupy poorer sandy soils. Alders and willows tend to grow in poorly drained, wetter soils. The deciduous forest of North America, to the south of the Great Lakes, tends to have a greater variety of plant species than the European forest. As the ice sheets advanced during the Ice Age tree species tended to migrate southwards. Subsequently, as the ice sheets retreated, a return migration to the north took place. In southern Europe mountain ranges trend generally from east to west and these interfered with the return migration of tree species. In North America, mountain ranges such as the Appalachians and the Rockies trend from north to south, so there was little interference with the migration of tree species. In southern and eastern USA the forest is 'mixed' (Fig. 3.2) rather than 'deciduous'. It contains broad leaved evergreens and also coniferous trees.

Mediterranean evergreens invading an abandoned quarry

Tropical rain forest (Congo)

Sparse coniferous evergreens in the dry interior of British Columbia

Boreal forest on the coast of British Columbia

Mediterranean woodland

This type of vegetation tends to occur on the west coasts of land masses between about latitudes 30° and 40° north and south, in California, central Chile, the southern tip of Africa, south-west and south-east Australia and on the coastlands, in particular, of the Mediterranean Sea. In these areas the summer season is hot (approaching 30°C) and the winter is comparatively warm (usually over 10°C). Most rain falls in winter and there is usually a very marked dry season in summer (Fig. 3.3(*e*)). A severe moisture deficit tends to occur in summer but there tends to be a surplus in winter. Often the most favourable growing seasons for vegetation are spring and autumn.

Along the coasts of the Mediterranean the true climax vegetation is difficult to identify. In mountainous areas, such as Corsica, forests of pine and fir are common. Altitude tends to reduce temperatures and the dry season is not so well marked as on the coastal lowlands. On lower ground there is much scrub vegetation. This mostly consists of evergreen shrubs with leathery, drought-resistant leaves. This dense scrub is referred to as *maquis*. The variety that grows in limestone areas is called *garrigue*. The Mediterranean coastlands have had a long history of human occupation and it is possible that maquis and garrigue have replaced forests which have been cleared by man. In some areas, however, it is possible that maquis and garrigue are the climatic climax vegetation, particularly in areas that are very arid, partly because of the climate and partly because of soil permeability. Scrub vegetation is also common in California, where it is called *chaparral*, central Chile and South Africa. In Australia there are eucalyptus forests in wetter areas and a scrub vegetation known as *mallee* in drier areas.

Temperate rain forest

This is found near the southern ends of South America, Africa, and Australia (Fig. 3.2). All these areas have an equable climate with very little frost and high annual precipitation totals. The climate illustrated in Figure 3.3(*f*) refers to southern Chile. Summer temperatures are comparable with those of the deciduous forest graph (Fig. 3.3(*d*)) but winter temperatures are considerably higher and rainfall totals are heavier.

In such a climate there is no need for trees to shed their leaves in winter since there is little frost. The forest is composed of a mixture of broad-leaved evergreens and conifers. Much of this forest has been cleared in southern Africa and Australia. Originally the Australian forest was composed of eucalypt hardwoods, and the kauri pine occurred in New Zealand.

SAVANNA

Savanna vegetation consists of a mixture of trees and grasses. Large areas occur in Africa, flanking the tropical rain forests. It is also found in South America in the campos of the Brazil Plateau and the llanos of the Orinoco Basin in Venezuela. It also exists on the northern and eastern margins of the desert in Australia (Fig. 3.2).

Savanna vegetation in Africa and Australia occupies the zone between the tropical rain forest and the desert. It therefore occurs in a wide variety of different climates. Temperatures are high throughout the year though there is a greater seasonal variation than in equatorial regions. In general, precipitation decreases from the forest margin towards the desert. Also, the length of the rainy season decreases towards the desert. The rainy season occurs in summer.

The traditional view is that savanna vegetation is a response to climatic conditions. In northern Africa various types of savanna have been identified which are arranged in broadly latitudinal belts, and these have been related to the progressive decrease in both annual precipitation and the length of the rainy season from the northern margin of the tropical rain forest to the southern edge of the Sahara Desert. At the edge of the forest is the 'high grass-low tree' savanna with grass (elephant grass) growing up to a height of 4 m. This contains many species of evergreen trees. Moving northwards, the grasses generally decrease in height and the proportion of deciduous trees tends to increase in response to the increasing length of the dry season. The next zone is the 'acacia-tall grass' savanna, the acacia tree being deciduous but some evergreen species still occur. The grass here tends not to be as tall as in the previous zone. This zone is succeeded by 'acacia-desert grass' savanna which reaches the southern edge of the Sahara. Here, the grass in much shorter and tufted and there is much bare soil. It is clear, therefore, that there are widely contrasting types of savanna and that it is not possible to identify a type of climate that is typical of a savanna vegetation. Wide variations in both temperature and rainfall regime exist.

More recently the savanna has come to be regarded as a plagioclimax vegetation (page 140) rather than climatic climax. The regular occurrence of fires could transform a forest vegetation into savanna. At the present time, fires regularly occur in the African savanna in the dry season. Fires would destroy most forest trees. It is significant that savanna trees such as palms and particularly the baobab with its thick bark tend to be resistant to fire. Grasses too can tolerate burning when they wither in the dry season and develop new growth in the succeeding rainy season. These fires may be started by man or they may be caused by lightning. Very frequently, there tends to be a sharp boundary between the rain forest and the savanna which could not be the result of a variation in climate. It is also significant that Africa has the largest areas of savanna and has also been occupied longest by man. It is also possible that soil conditions could account for the existence of some types of savanna. A waterlogged area, for example, might be occupied by grasses rather than forest trees. Also, grazing animals could help to prevent the growth of trees without interfering with grass growth.

TEMPERATE GRASSLAND

Very few areas of temperate grassland still exist. In most areas they have been developed as farming land and the original extent of the grassland is not fully known. The same problems arise here as were outlined in respect of savanna vegetation.

Figure 3.2 shows the distribution of the areas that are generally believed to have been occupied by temperate grassland. Like the savanna, they tend to occur in the transitional area between

Savanna (The Golden Gate area of South Africa)

forested areas and arid lands. In North America the temperate grassland extends from the Canadian prairies to the Gulf of Mexico, between the mixed forests of eastern USA and the arid lands of the west. In Eurasia grasslands occur similarly between forest and desert, but here their orientation is east-west rather than north-south. A similar pattern exists in the southern hemisphere in respect of the Argentine pampas, the South African veld, and the Murray-Darling Basin in Australia. In New Zealand the Canterbury Plains lie in the rain shadow of the Southern Alps.

Most of these grasslands have a relatively light annual rainfall, often with a maximum in early summer (Fig. 3.3 (*g*)). In the northern hemisphere, their continental location and relatively high latitude give the Canadian prairies and the Eurasian steppes particularly cold winters, but summers are comparatively warm. In natural grassland areas, growth can begin in spring and seeding can take place before the succeeding winter. Growth is aided by the early summer rains. In the southern hemisphere grasslands, temperatures are considerably higher in the winter season.

The natural vegetation appears to consist of tall grasses with some trees in the wetter areas, grading to short grass and eventually scrub at the arid margin. This progression is from east to west in North America and from north to south in Eurasia. Trees can exist in the drier areas, particularly along the courses of rivers. However, as in the case of the savanna vegetation (page 147), it is by no means certain that the grassland-forest boundary reflects the influence of climate alone. Before Europeans arrived in North America the Indians had been accustomed to firing the grassland, and also grazing by bison could have prevented trees from invading the grassland.

DESERT AND SEMI-DESERT VEGETATION

An area of desert is found in all parts of the world where either the Tropic of Cancer or the Tropic of Capricorn crosses the west coast of a major land mass. The greatest area of desert extends from north-west Africa across northern Africa and the Middle East to north China (Fig. 3.2). Desert exists in a similar location in the south-west of North America. Much of Australia is desert as a result of its position astride the Tropic of Capricorn. In South America and southern Africa deserts are generally restricted to fairly narrow coastal strips.

There are great extremes of temperature particularly in the larger deserts. A location in the Sahara has a range of monthly mean temperatures from 36°C to 13°C and practically no precipitation. In contrast, some coastal deserts are much more equable and moisture may be available from mists. In southern Africa, for example, Walvis Bay and Port Nolloth have remarkably equable temperatures (Fig. 2.8).

In general there are two contrasting types of vegetation in desert areas. Some plants are in a dormant state for most of the time and become active following the occurrence of rain. At these times such plants hurry through a rapid life cycle and produce seed protected from desiccation by either the seed coating or the remains of the plant. Others in this general group are able to store water from the occasional rains in either succulent leaves and stems or underground tubers. Such plants as these appear to represent a climatic climax vegetation since they are directly adapted to the climate. The other group of plants is not so directly dependent on rainfall. These obtain their moisture from the water table. Such plants are often distributed along depressions such as the floors of wadis where their long roots can reach the water

Vegetation in an arid climate (Tenerife)

table. They are perennial plants that counter the desiccating effect of the desert climate by reducing transpiration. This is achieved by having very small leaves or spines, the leaves having thick cuticles and often a cover of wax or hair.

TUNDRA AND ALPINE VEGETATION

Tundra vegetation is more or less restricted to the northernmost parts of North America and Eurasia, together with the coast of Greenland and a number of Arctic islands (Fig. 3.2). Generally it forms a ring around the Arctic Ocean. There is little tundra in the southern hemisphere because these latitudes are mainly occupied by ocean. The Antarctic continent is so far south that it is generally too cold to support tundra vegetation. In Eurasia, particularly, tundra vegetation tends to extend southwards along mountain ranges (Fig. 3.2).

Areas of tundra are not always excessively cold in winter but their summers have mean temperatures below 10°C (Fig. 3.3). In winter strong winds tend to desiccate the vegetation even though temperatures are not excessively low. The summer frost-free season lasts less than two months. Despite the long hours of daylight in the summer, the midday sun only reaches a low elevation in the sky (Fig. 2.3) and this keeps temperatures low.

There are no large trees in the tundra because the soil above the permafrost is comparatively shallow, winds are often excessively strong and the growing season is so short. Some plants assume a cushion form, like the saxifrages (which can be seen in rockeries in Britain). Others die down in winter and survive as corms. Some annuals manage to produce seed in readiness for growth in the succeeding summer. The type of vegetation is mainly influenced by three factors: drainage conditions, the degree of exposure to desiccating winds, and aspect. Most of the tundra is poorly drained because of the underlying impermeable permafrost. The shallow soil layer is wet throughout the summer. In such areas cotton grass, various marsh plants and the almost universal mosses and lichens grow. This vegetation tends to build up layers of peat. Very little vegetation exists on coasts that are exposed to the wind. Strong winds and snow driven by the wind exclude growth unless some shelter is available. Inland, especially near the southern margin of the tundra, it is common to find stunted trees of alder or birch in the valleys and on sheltered hill slopes, with low growing plants on exposed summits. Some south-facing hill slopes may have deeper, drier soils and higher summer temperatures as a result of more effective insolation. Here, there may exist a more widespread cover of flowering plants such as anemones, buttercups and primroses.

Alpine vegetation occurs on mountains in various parts of the world, between the upper limit of tree growth (the tree line) and the snow line. Its height progressively decreases from the equator towards higher latitudes. Many plant species of the tundra also occur in this highland environment. The progressive decrease of temperatures with increasing latitude is comparable with the fall in temperature as altitude increases. But the mountain environment differs in several ways from that of the tundra even though mean annual temperatures may be similar. On the equator, for example, there is little annual variation of temperature whereas the tundra can have a considerable range. But at high altitudes, with a thin atmosphere, there may be intense heating of the surface during the day and intense cooling at night. There may also be great differences in temperature between sunlit slopes and shaded slopes. In mountainous areas of temperate and tropical regions summer days are shorter but generally warmer than in the tundra and some degree of insolation can occur throughout the winter, whereas in much of the tundra there is a period when the sun never rises. Mosses and lichens are common in highland areas, as in the tundra, particularly on steep rock slopes and screes where there is little soil. Also, cushion-shaped plants such as saxifrage are common. Alpine areas also have flowering plants pushing through snowdrifts in spring, rather like those of the tundra, followed by other plants which succeed in forming their seed before winter comes.

Exercises

1. Illustrating your answer by means of a 'system' diagram (in the same style as Fig. 1.14(*c*)), explain what is meant by the term 'ecosystem'
2. Referring to specific examples, explain why the life forms of the vegetation may be similar in different parts of the world that have a similar climate but, on the other hand, the plant species may be quite different.
3. (*a*) Explain the terms 'climatic climax', 'edaphic climax', 'prisere' and 'plagiosere'.
   (*b*) Why is it rather rare to find examples of true climatic climax vegetation?
4. Discuss the problems that are involved in interpreting the vegetation of grassland and savanna areas.
5. In what ways are the characteristics of various types of vegetation influenced by: (*a*) annual precipitation totals; (*b*) the seasonal distribution of precipitation?
6. Discuss the factors that have influenced the distribution and characteristics of evergreen and deciduous forests.
7. Describe and explain the changes in the type of vegetation that you would expect to occur along each of the lines A–B in Figure 3.4 (*a*) to (*d*).
8. Figure 3.5 shows the characteristics of examples of five different climatic types (A to E). For each of these
   (*a*) suggest a related type of vegetation;
   (*b*) explain the relationships between the climate and the vegetation.

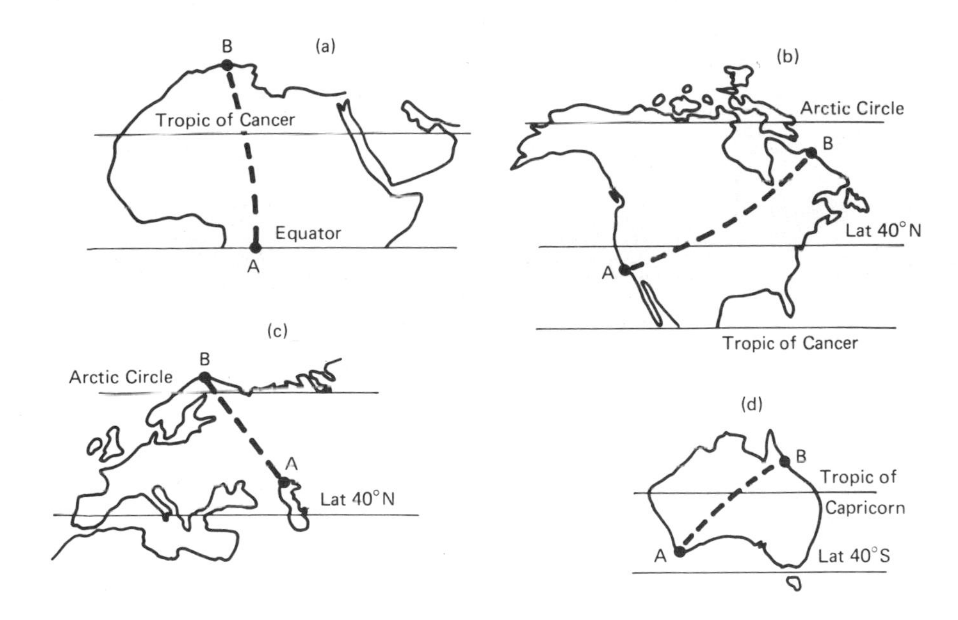

Fig. 3.4

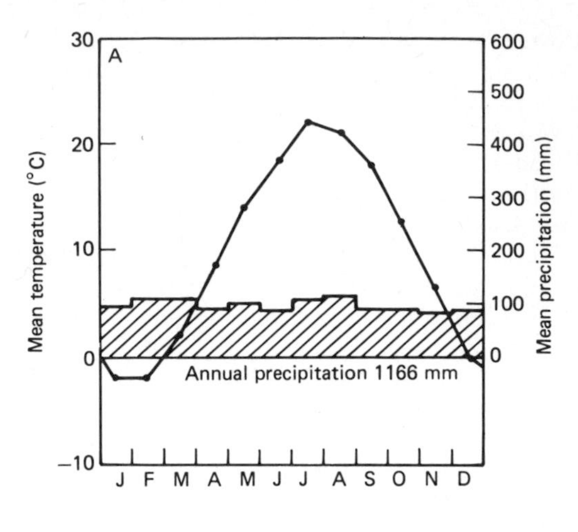

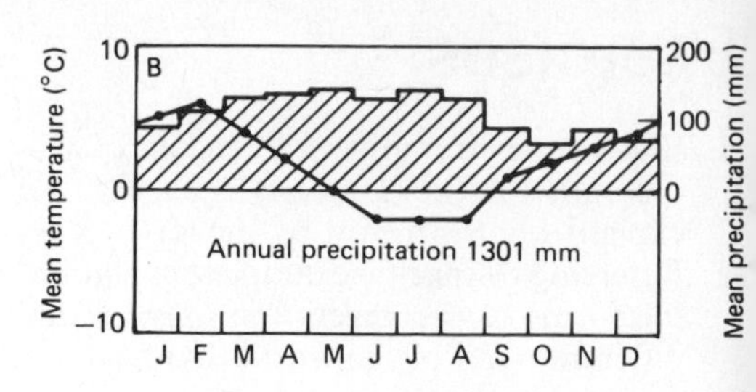

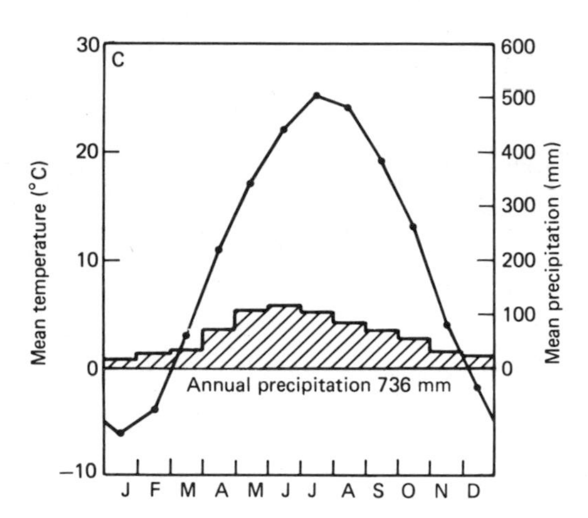

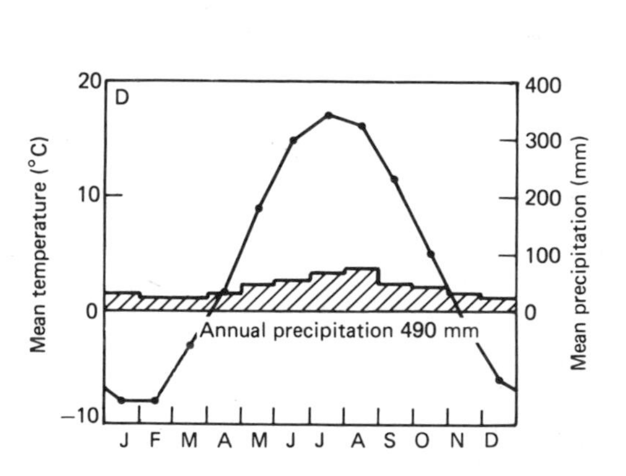

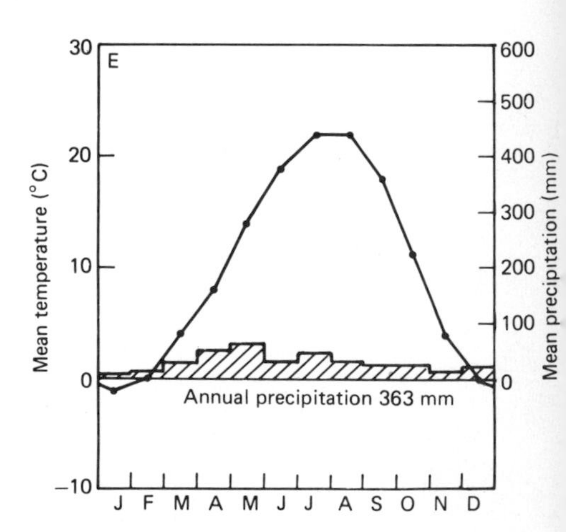

Fig. 3.5 Five different climatic types

4 Soil

4.1 The place of soil in an environmental system

It is fitting that the last section of this book should be devoted to soil. Thc soil can be regarded as an important part of an environmental system in which interaction takes place between living and non-living elements (Fig. 4.1). The soil is an essential link between organic and inorganic elements of the environment.

Much of the material in soils is derived from rocks. By the various processes of weathering, coherent rock is reduced to regolith, which consists of particles of mineral material (pages 7–10). This is not soil, but it constitutes the parent material of soil. The particles can be of various sizes, from pebbles to sand and silt to fine-grained clay. Organic material derived from plants or animals is added to this rock-waste in the form of decayed leaves or roots or waste products from animals. This is referred to as *humus*. Thus, a soil develops which is capable of holding moisture within itself against the force of gravity. Warmth and moisture to assist in this process of organic decay come from the atmosphere and through the atmosphere by direct insolation (Fig. 4.1). Moisture is also evaporated from the soil into the

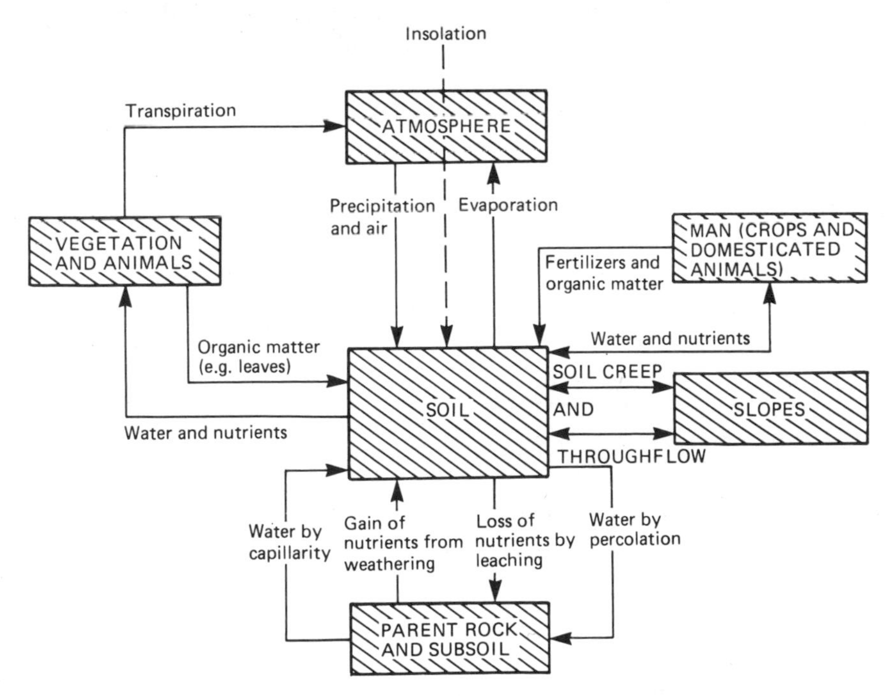

Fig. 4.1 Soil as an element in an environmental system

atmosphere and the relationship between precipitation and evaporation determines to some extent, the type of soil that eventually develops. Soil water and various nutrients (plant foods) in the soil support the growth of vegetation and this vegetation supplies food for animals. Some of this moisture is eventually transpired into the atmosphere by vegetation.

Water may pass downwards through the soil into the subsoil and the parent rock, especially if precipitation is greater than evapotranspiration. This process can cause the soil to lose some of its nutrients by leaching. In other cases weathering of the parent rock can provide extra plant nutrients and deepen the soil. Under certain climatic conditions it is possible for soil water to move upwards by capillarity (Fig. 4.1) so that plant nutrients may be added to the soil.

Soil development may also be influenced by certain site characteristics, such as location in relation to hill slopes. On steep hill slopes soil may be removed by mass movements (pages 10–12) but other sites may accumulate soil that has been transported from higher parts of the hill slope (Fig. 1.10). Similarly, moisture may be lost from the soil by throughflow in certain sloping sites (page 17) but soil water may accumulate in other locations.

Finally, the activities of man may influence the characteristics of the soil. Water and nutrients in the soil support crops and domesticated animals. Also, farming activities may involve the return of organic matter to the soil and the renewal of soil nutrients by the application of artificial fertilizers.

4.2 Soil characteristics and processes

It is clear from the preceding section that the characteristics of soil are quite different from those of the underlying rock, and that certain processes occur in soil that normally do not occur in the underlying rock. The special importance of soil lies in the fact that it is a link between the lifeless world of rock and rock waste and the living world of plants, animals and man.

SOIL CHARACTERISTICS

The composition of soil

Soil contains solid materials of both organic and inorganic origin and also various liquids and gases. Soil water (or soil solution) exists in the pore spaces of the soil and can move through the soil. It is supplied to the soil by precipitation and it can move out of the soil, by evaporation, by transpiration through plants, by throughflow to river channels, and by percolation downwards towards the water table (Figs. 1.14(*b*) and (*c*)). Most of the processes that take place in soils are dependent on the presence of soil water.

Air also enters the soil from the atmosphere and occupies the pore spaces that are not occupied by water. Air can only enter the soil if it is not fully saturated with water. Certain soil processes can only take place if air is present. For example many types of bacteria can only exist where air is present and bacteria are largely responsible for the decomposition processes by which plant and animal material is transformed into humus. Soil air usually has less oxygen and nitrogen, but more carbon dioxide, than the atmosphere.

Solid material in soil may be organic or inorganic in origin. The inorganic part of the soil consists of weathered material derived from the parent rock. This may consist of sand grains (silica) weathered directly from the parent rock with no chemical change. However, clay in the soil may be the result of chemical weathering of aluminium silicate minerals (page 8). Weathering of the parent rock also supplies the soil with compounds of sodium, potassium, calcium and magnesium (bases) which are important nutrients for plants.

The organic part of the soil consists of dark brown or black humus. This is organic material, derived from plants and animals, that has been partly decomposed by bacteria, earthworms and fungi. The stems and leaves of plants are first deposited on the soil surface and then mixed into the soil by various organisms. Decaying plant roots, on the other hand, are already incorporated in the soil. The amount of humus present in a soil depends primarily on the amount of vegetation that the soil supports, but it also depends to some extent on the climate. High temperatures favour the rapid production of humus. The type of humus present in a soil largely depends on the type of vegetation that the soil supports. Some plants and trees take up a large quantity of nutrients from the soil and use these to build up stems, leaves and root systems. These are eventually returned to the soil as mull humus that is rich in plant nutrients (bases). This kind of cycle is particularly characteristic of grassland and deciduous woodland. Other types of vegetation, such as coniferous evergreen trees, take up fewer plant nutrients from the soil. Hence, they return to the soil a much poorer type of humus known as mor. The presence of humus tends to give the soil a brown or black colour. Iron compounds, on the other hand, give red or yellow colours. But in soils which are poorly drained, iron compounds suffer reduction because of the lack of oxygen and give the soil a green, grey or blue colour.

Soil textures

The term 'texture' refers to the size of the individual particles that make up a soil. Three major groups of particle sizes are identified. These are, in order of decreasing size, sand, silt and clay (Fig. 4.2). A particular soil may be a mixture of all these particle sizes. Such a soil is termed a *loam*. Other mixtures are termed clay loam, silt loam, and sandy loam. Figure 4.2 illustrates possible types of soil texture. The texture of a soil determines, to a great extent, its permeability. Water percolates very rapidly through a sandy soil but a clay soil tends to hold water in its tiny pore spaces. A clay soil tends to feel sticky when it is wet, a silty soil feels smooth, and in a sandy soil the individual sand grains can be felt. Soil textures are related to the parent rock from which the soil evolved. Sandstones weather to produce sandy material, shales produce silty material. Acid igneous rocks which are rich in quartz (such as granite) tend to produce sandy material, whereas basic igneous rocks such as basalt tend to produce clay. The permeability of a soil is also influenced by pore spaces created by worms or burrowing animals and the decay of plant roots. Soils also contain tiny particles called *colloids* that may be composed of either mineral or organic material.

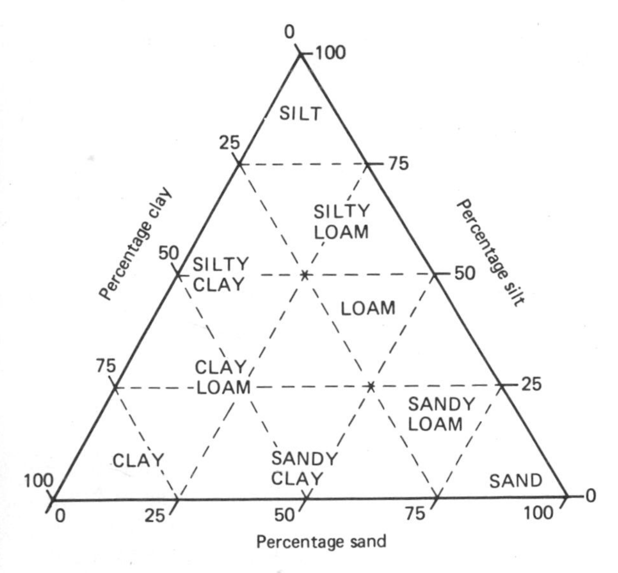

Fig. 4.2 Soil textures

Soil structure

The term 'structure' refers to the way in which grains of soil adhere together to form larger pieces which are referred to as *peds*. The soil grains are held together by colloids. Various structures are identified in relation to the size and shape of the peds. In granular (or crumb) structure, soil grains adhere together to form rounded pieces. Larger aggregations of irregular shape, with sharp corners, are characteristic of blocky structure. Columnar structure is characterized by vertically elongated, prismatic columns. In platy structure, on the other hand, the soil grains form flattish, horizontal aggregations.

The clay-humus complex

Soil does not simply consist of humus mixed with particles of sand, silt and clay. Humus tends to be linked with clay particles to form the clay-humus complex. Here, tiny particles (*micelles*) of clay and humus can hold plant nutrients (bases) in such a way that they are available to the roots of plants. If the clay-humus complex were not present, rain water would be able to wash soluble plant nutrients downwards through the soil.

Soil acidity

Some soils are naturally alkaline. Examples are recently developed soils on chalk or limestone (calcium carbonate) surfaces. But many soils tend to be acid because the bases of the clay-humus complex tend to be replaced by hydrogen from soil water. The pH value of a soil is a measure of the concentration of hydrogen ions in the soil water. A pH value of 7 is neutral. A value greater than 7 indicates an alkaline soil and a value less than 7 indicates an acid soil. A slightly acid soil is usually

A boulder clay soil

A boulder clay soil under a grassland vegetation

best for plant growth since many soil nutrients are then most soluble and easily available to plants.

MAJOR SOIL PROCESSES

Cation exchange

The clay-humus complex is fundamental to the various processes that take place in soils. It holds the largest reserve of plant nutrients in the form of the cations (bases): calcium, magnesium, potassium and sodium. Grains of purely mineral sand and silt can supply no nutrients to plants. Soil colloids are electrically charged and can attract base cations from the soil solution. These nutrients come to be attached to the micelles of the clay-humus complex. If the clay-humus complex did not exist these bases would be washed (*leached*) out of the soil by percolating water.

When the roots of plants come into contact with the clay-humus micelles they give off hydrogen ions in exchange for plant nutrients. Thus, the bases of the clay-humus complex are replaced by hydrogen ions. The bases can subsequently be renewed by the creation of new supplies of humus by the recycling of plant debris (page 155).

Cation exchange also takes place between the clay-humus complex and the soil solution.

Hydrogen ions from the soil solution can replace the base cations on the surfaces of the soil colloids, thus causing an increase in the level of soil acidity. When the clay-humus complex is almost completely saturated with hydrogen the soil has a pH value of about 4. The clay-humus complex is stable when it is saturated with base cations. As these are replaced by hydrogen ions it tends to become increasingly acid and unstable. Eventually the clay-humus complex may disintegrate and be washed downwards (leached), in an extreme case leaving a purely mineral soil. The application of fertilizers can have the effect of renewing the cations of the clay-humus complex and preventing a decline in soil fertility.

The soil profile

The term 'soil profile' refers to the variations that occur in the characteristics of a soil downwards from the ground surface to the underlying parent rock. It is sometimes called the *solum* (Fig. 4.3). A well-developed soil tends to have horizontal layers (horizons) which differ in texture, colour and chemical composition. These horizons are transitional between mineral material at a depth and organic material at the surface. They are created by processes that operate within the soil.

In the surface layers organic matter, mostly derived from the vegetation, tends to accumulate in the A_O or O horizon (Fig. 4.3). This can commonly be divided into three layers which are referred to as L (litter), F (fermentation) and H (humus). They represent three stages in the decomposition of organic material. The L layer consists of loose leaves and other raw organic debris. Beneath this, in the F layer, active decay is taking place and the organic debris has partly decomposed. In the underlying H layer decay is more complete and the organic material has become dark coloured humus which combines with the clay fraction of the soil to form the clay-humus complex.

Beneath these organic layers, in the A and B horizons, the mineral content of the soil increases downwards towards the parent rock from which it is derived.

In a humid temperate climate the soil solution generally moves downwards through the soil since precipitation is greater than evapotranspiration. In the A horizon leaching tends to take place, and soluble soil constituents are carried downwards. Hydrogen ions from the soil solution replace base cations in the clay-humus complex (page 155) and the cations are carried away, thus increasing the soil's acidity. The clay-humus complex may possibly disintegrate and be leached downwards out of the A horizon.

The underlying B horizon generally has a smaller humus content and a greater proportion of weathered rock material, but substances leached from the A horizon may accumulate here. Only the A and B horizons are composed of true soil. The C horizon consists mostly of weathered parent rock, and the D horizon is unchanged parent rock.

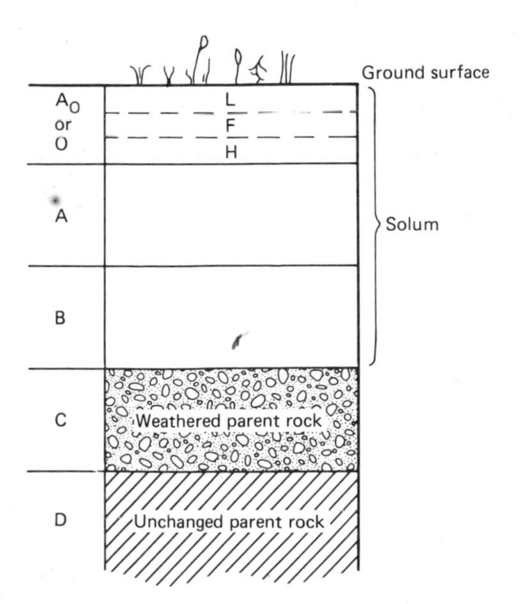

Fig. 4.3 Divisions of a soil profile

4.3 The classification of soils

In various kinds of studies in physical geography it is convenient and useful to use systems of classification. In studying drainage patterns we classify both the types of streams and the drainage patterns themselves (pages 27–32). We have also used classifications for types of climate (pages 125–132) and types of vegetation (pages 142–150). This is necessary in order to organize our knowledge systematically. By developing classification systems we can come to understand broad generalizations in addition to detailed individual descriptions or explanations.

In some cases a classification is made up of a hierarchy of elements. Major categories are identified in terms of major characteristics, then each of these is divided into sub-groups in respect of more detailed characteristics. Classifications do not add to knowledge. Their value lies in the fact that they organize knowledge more efficiently even though, at times, anomalies may appear.

A classification of soil types that has proved useful in geographical studies is the zonal system. In this system the world's soils are subdivided into three major categories, zonal, intrazonal and azonal. *Zonal* soils have characteristics that tend to reflect climatic conditions. *Intrazonal* soils tend to reflect other factors such as the characteristics of the parent rock. *Azonal* soils are generally immature and skeletal, with poorly developed soil profiles.

ZONAL SOILS

A zonal soil is one that has been created over a long period of time by the influence of climate and the associated vegetation. The ground surface under which it develops is gently undulating and the soil is well drained. The parent rock material is not of a special type such as limestone that would have a strong influence upon the nature of the soil. In these conditions, climate is the main factor that influences the soil's development. This concept of a zonal soil has attracted much criticism. It has been argued that considerable climatic changes have taken place during the period in which zonal soils have evolved and, in any case, there can be many small-scale climatic variations within quite a small area. Soil drainage conditions can also vary considerably. The concept of a zonal soil is therefore rather similar to that of the climatic climax vegetation (page 138). Nevertheless, the zonal soil concept has made it possible to describe and explain in relatively simple terms the broad features of the distribution of soil types over the world and to relate them to the climatic climax vegetation. It is not possible to study local, small-scale soil variations in terms of zonal soils.

Tundra soil

The distribution of areas of tundra vegetation is shown in Figure 3.2. In the tundra areas the subsoil is permanently frozen (permafrost), summers are very cool, and the growing season is short. The soil is frozen in winter and waterlogged in summer since water cannot percolate downwards because of the frozen subsoil. Most of the mineral material consists of frost-shattered rock fragments contained in a blue-grey mud in which ferric oxides have been reduced to blue or grey ferrous salts in the absence of oxygen in the soil because of the waterlogging (gleying). The surface layer often consists of peat. This exists because bacterial activity is so weak and therefore the development of humus is very slow. Unlike most zonal soils there are no clear soil horizons. This is because alternate freezing and thawing causes much disturbance, and the underlying mud is sometimes forced to the surface through the peat layer.

Podzol

The podzol is the zonal soil of the boreal forest area (page 144, Fig. 3.2). It is also found in other areas that have cool climates, such as the United Kingdom. Here, it occurs on heaths and moors and in areas of sandy soil, such as outwash plains created by fluvioglacial processes in the Ice Age. Podzols are often associated with coniferous evergreen trees, particularly pines.

Areas of boreal forest do not receive a particularly heavy rainfall but the podzolization process requires a general downward movement of water through the soil. Precipitation is greater than evapotranspiration, and evapotranspiration is at quite a low level because of the generally low temperatures throughout the year. Also, the nature of the forest vegetation influences the moisture balance. Coniferous evergreen forest vegetation casts a heavy shade over the ground surface and also shelters it from winds. Hence, there is little evaporation at the level of the forest floor. A moderate amount of precipitation can therefore provide a surplus over evapotranspiration and soil water can percolate downwards through the soil.

The podzol has a very poor nutrient cycle. Coniferous evergreens, particularly pines, do not take up bases such as calcium, magnesium and potassium, so such bases are not returned to the soil surface when the leaves fall. The humus produced is therefore of the mor (acid) type. The A_0 horizon of a podzol consists of relatively unaltered leaves at the surface (L horizon, Fig. 4.4(*a*)) and these are increasingly decomposed downwards through the F and H horizons. The A_1

horizon is usually dark grey in colour just below the H horizon, because of staining by humus, but it becomes lighter in colour downwards and grades into the A_2 horizon which is composed of a light grey (sometimes almost white) sandy material. All colouring matter, such as humus and iron and aluminium sesquioxides, has been leached out of this horizon. Also, bases and colloids have been leached. At the base of the A_2 horizon there is a sudden transition to the B horizon which is much darker in colour and denser in texture. Here, clay, humus, and sesquioxides have been deposited. In the B horizon iron may accumulate to form a hardpan (ortstein). This commonly occurs near the highest position reached by the water table. Lower down, the B horizon merges into the parent rock. The various horizons in a podzol tend to be very strongly marked, with sharp junctions. This is because there are so few mixing agents such as worms.

Brown forest soil

The climate associated with this type of soil is wetter and warmer and provides a longer growing season than that of the boreal forest. The vegetation here is deciduous woodland (page 144, Fig. 3.2). The deciduous trees supply the soil with more nutrients than in the case of the podzol. Some deciduous trees, including the oak, send their roots very deep and draw nutrients from deep in the soil and even from the parent rock. Thus, there is a constant supply of plant nutrients and these are returned to the soil surface when the leaves are shed annually. Soil fauna are more abundant than in a podzol so plant debris is broken down efficiently, and thoroughly mixed into the soil by earthworms and rodents, and bacterial activity is more intense. Hence, the brown forest soil contains more bases and humus than a podzol. Generally, therefore, leaching is not so severe but it can increase if the parent rock has few bases or if the precipitation is particularly heavy.

Some leaching, however, does naturally occur and the upper part of the profile tends to be slightly acid. Clay and humus tend to be leached downwards but there is not usually a clear differentiation between different soil horizons.

The A_0 layer consists of a litter of leaves underlain by a layer of dark brown, slightly acid humus that is rich in plant nutrients. The remainder of the profile down to the parent rock gradually becomes a paler brown in colour. This is because humus tends to become less abundant with greater depth.

Chernozems and related soils

The chernozem soil is associated with temperate grassland vegetation (Fig. 3.2). The climate here is generally warmer and drier than in the forest areas, though, in the continental interiors, winters may still be very cold with much frost. In these grassland areas moisture from the soil evaporates freely from the ground surface and much soil moisture can be lost through transpiration by the grasses. Rainfall, therefore, does not usually penetrate deeply into the soil and sometimes soil water can move upwards through the soil. In addition, the soil usually freezes in winter and this tends to

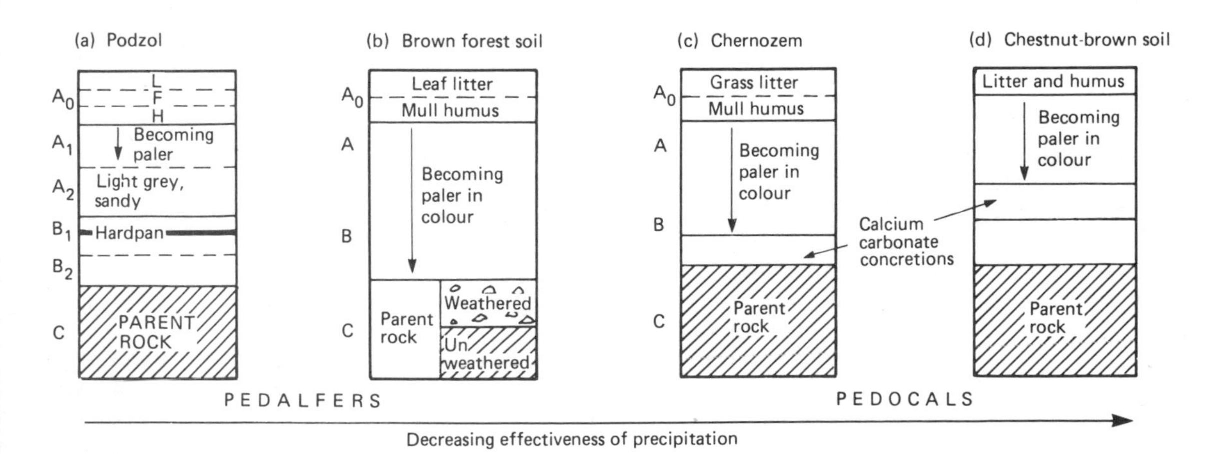

Fig. 4.4 Zonal soils of forests and grasslands

prevent leaching. The chernozem has a very rich nutrient cycle. The grassland vegetation is able to take up more nutrients from the soil than a forest vegetation. Thus, the humus that is returned to the soil is richer than that received by forest soils.

The chernozem has a particularly rich soil fauna so that humus from the A_0 horizon (Fig. 4.4) is quickly mixed into the soil. This tends to give a dark brown colour to the upper part of the A horizon. From here the colour of the soil gradually lightens downwards towards a yellowish brown colour. A chernozem contains a considerable amount of calcium carbonate. This tends to be deposited in the burrows of the soil fauna and it also tends to form a layer of nodules at the maximum depth to which rainfall penetrates (Fig. 4.4). The presence of calcium carbonate in the soil counteracts any tendency for clay compounds to break up and for leaching to begin.

Prairie soils are found in areas that receive rather more precipitation than those in which chernozem soils occur, but they are still associated with a temperate grassland vegetation. They can be regarded as transitional between the brown forest soil and the chernozem. In the prairie soil there is no layer of calcium carbonate concretions as there is in the chernozem. All independently occurring calcium carbonate has been leached. However, the grassland vegetation is demanding in respect of soil nutrients and, when it decays, it returns a rich supply of nutrients and humus to the soil. Although the soil is slightly acid, very little leaching occurs. There is little tendency for contrasting horizons to occur in the soil profile. A and B horizons merge into a single dark brown layer.

Chestnut-brown soils are found on the drier side of the chernozem soils under a vegetation of rather poorer grassland. They therefore contain less humus. They are brown in colour and become paler downwards through the profile. Because of the dry climate there is hardly any leaching. As in a chernozem, a chestnut-brown soil has a layer of calcareous concretions, but it is at a shallower depth than in a chernozem (Fig. 4.4(*d*)).

On the drier side of the chestnut-brown soils, on the desert margins, the zonal soil is the *serozem*, a greyish soil with a very low humus content. Vegetation in these areas is very sparse. The grey colour is partly due to the high calcareous content brought towards the surface by the upward movement of soil water. This soil is rich in mineral nutrients and is fertile if irrigated.

Tropical red soils

These soils can be formed under a tropical rain forest vegetation (page 142) where drainage is good. The tropical rain forest vegetation supports a rich nutrient cycle, returning to the soil a constant supply of the nutrient bases that have supported forest growth. Decomposition of the leaf litter is very rapid in the hot, moist climate. Because of this, the soil tends not to be very acid. It appears, however, that under these conditions silica can be leached downwards leaving sesquioxides of iron and aluminium in the surface soil. These become more concentrated and ultimately a lateritic crust may develop on the surface. In this process clay colloids have been removed by leaching. Iron compounds give the soil a red colour. Aluminium oxide can also become abundant in the soil. If the forest vegetation still exists above such a soil it is thought that it survives largely through the nutrients contained in the surface soil that are derived from the leaf litter. It appears that lateritic soils do not develop fully when there is a cover of rain forest.

Pedocals and pedalfers

Zonal soils may be subdivided into two groups known as pedalfers and pedocals. Pedalfers are soils in which iron and aluminium oxides tend to accumulate. Pedocals are those in which calcium carbonate tends to accumulate. Pedocals generally occur in drier climates where leaching is less intense. They include chernozems and chestnut-brown soils, both of which have a layer of calcium carbonate concretions (Fig. 4.4). Pedalfers occur where precipitation is heavier and leaching is more intense. They are characterized by a layer composed of iron and aluminium sesquioxides. This can exist for two quite different reasons. In podzols it is because the sesquioxides have been leached downwards into the B horizon where they can form a hardpan (page 159, Fig. 4.4(*a*)). In tropical red soils, however, the concentration of sesquioxides is a residual feature near the surface and is the result of the leaching of silica.

INTRAZONAL SOILS

The characteristics of intrazonal soils are not primarily the result of climatic influences. Within any climatic zone there can be great variations in relief, rock type and drainage conditions and these can have a great influence on soil type. A limestone subsoil, for example, can give rise to the occurrence of alkaline soils in areas where the climate might be expected to produce leached, podzolic soils. Also, it is possible for a particular type of intrazonal soil to occur in different regions which have different climates. Three main types of intrazonal soil are distinguished according to the predominant influence upon their characteristics. These are the hydromorphic type that is developed in areas where there is an excessive amount of water, the calcimorphic type that is associated with a limestone subsoil, and the halomorphic type which consists of soils with a high salt content.

Hydromorphic soils

Hydromorphic soils usually develop in swampy areas (bog soils) and in lowlying river flood plains (meadow soils). In these soils part of the profile is constantly saturated with water and the level of the water table can vary from season to season so that even the upper parts of the soil are periodically saturated. Often large amounts of organic matter in the form of plant remains accumulate in such soils but the presence of large amounts of water tends to restrict the activity of bacteria that could transform this organic matter into humus. The result is that partly decomposed vegetation in the form of peat commonly occupies the upper horizon of these soils. Bog peat forms in poorly drained upland areas and is extremely acid. In lowland areas of fen, however, the presence of river water containing calcareous material can result in the formation of a less acid type of peat.

Lower levels of the soil are saturated for longer periods. Any oxygen is rapidly used up by micro-organisms and this results in gleying. Iron compounds in the soil are reduced chemically (give up their oxygen) from the ferric state to the ferrous state. This involves a change in colour from red or brown to green or blue. Other minerals in the soil can produce a grey colour. In places where oxygen is available from air or percolating water, such as along root channels or small fissures in the soil, oxidation takes place and produces red or brown mottles or streaks. These can exist when the rest of the soil is almost uniformly grey or blue. The depth of the gleyed soil horizon varies with the position of the water table, but gleying generally becomes more intense at greater soil depths.

Gleying is not restricted to the types of soil described above. Podzols can sometimes show mottling beneath the hardpan (page 159). It is also possible for gleying of the soil to take place much nearer the soil surface. Where the soil consists of boulder clay which is relatively impermeable, there may be no identifiable water table. In this case the soil may become saturated from time to time and gleying can occur at a very shallow depth. Such a soil is referred to as a surface water gley.

Calcimorphic soils

These soils occur in limestone areas. The *rendzina* is a dark-coloured soil, fairly rich in humus, that develops in humid climates on softer limestones such as chalk. There is usually a sharp transition between the soil and the underlying rock, marked only by a thin layer of fragments weathered from the limestone. The rendzina is sometimes called an A–C soil because of the absence of an illuvial B horizon in which leached materials accumulate. This is because there is little leaching of the A

A rendzina-type soil on the Marlborough Downs

horizon and any leaching that does occur involves soluble materials. Also, the underlying limestone, when it weathers, is mixed upwards into the upper horizon. The pH of a rendzina is over 7, indicating slight alkalinity.

On more resistant types of limestone weathering tends to be slower and hence upward mixing of weathered material is not so marked. The soil therefore tends to be more acid and the clay minerals in the soil can be broken up. This produces ferric oxide which tends to give the soil a reddish colour. Such a soil is referred to as a *terra rossa*.

Halomorphic soils

Halomorphic soils mostly occur in arid climates. As their name implies they are created by the concentration of various soluble salts in particular localities. The most extreme example is the *solonchak*. This has the greatest concentration of salts such as sodium chloride, gypsum and calcium carbonate. These may rise by capillarity, if the water table is at a fairly shallow depth, and form a crust on the surface. Alternatively, the salts may be carried in solution by surface run-off which collects in lowlying areas and may then be precipitated as the water evaporates. A solonchak can also form as a result of irrigation with insufficient drainage. Salts are precipitated from the irrigation water. A *solonetz* is a thin grey soil that overlies a dark-coloured alkaline layer. It has a smaller concentration of salts than a solonchak. This may be because the water table is lower, so that less salt rises by capillarity. Alternatively, it can result from greater leaching of salts in an area where the rainfall is greater. A third type is the *soloth* which is a degraded form of the solonetz. It is similar to a solonetz but it is more fully leached and has a lower salt content. This may be because of a higher rainfall or a greater supply of irrigation water combined with adequate drainage.

A solonetz

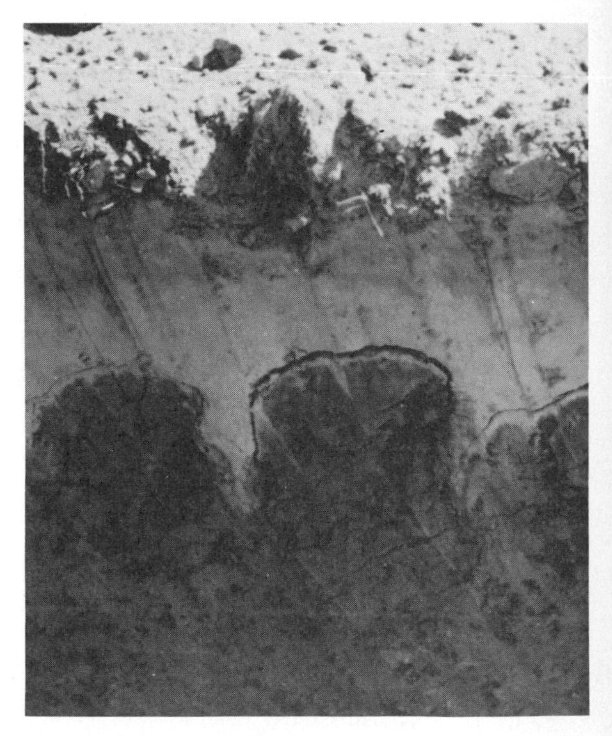

A podzol

AZONAL SOILS

Azonal soils are immature soils made up almost entirely of fragments of mineral material and usually containing very little humus. For various reasons they have been affected relatively little by the soil-forming processes described above that have created the characteristics of zonal and intrazonal soils. They tend not to have well-developed soil profiles. This may be because they have developed very recently. Azonal soils are usually classified into three groups. *Lithosols* are composed mainly of material of coarse calibre, such as sand or gravel that occurs in hilly areas in the form of screes or newly created glacial moraines. They often occur on slopes where surface run-off of precipitation is common. Thus, any fine-grained rock debris that is produced by weathering tends to be washed downslope. They are also affected by mass movement resulting from slope processes (page 10). Little vegetation is able to grow under these conditions so there can be little humus in such soils.

The second group, *fluvisols*, is made up of relatively recently deposited alluvium in and near the flood plains of rivers. Often a grass vegetation supplies some humus to these soils. In areas of flood plain there is a fairly high water table (Fig. 1.14(*b*)) and gleying (page 161) often occurs. Soils on nearby river terraces (page 46) are better drained and have existed for a longer time, so commonly they have developed some of the characteristics of zonal soils.

Regosols, the third group, are soils consisting of fairly deep layers of relatively fine rock material in, for example, sand dunes, wind-deposited loess and glacial and volcanic deposits.

A SOIL CATENA

A soil catena is a sequence of various types of soil profile that occur in a succession from the crest to the foot of a hill slope. The term 'catena' is usually restricted to hill slopes with a uniform type of underlying rock. The sequence of profiles is therefore related to the characteristics of the slope itself. Variations in soil type along the slope are mainly the result of variations in soil drainage which is faster on steep slopes than on gentle slopes.

If the summit of the slope is a plateau, most of the soils could be of the zonal type, directly related to the climate and vegetation, such as a podzol in a cool, moist climate. On a level surface drainage may be poor and the soil may have developed a gley horizon low down in the profile. With better drainage, soil nutrients may have been leached and carried downslope through the soil. These could accumulate in the soil near the base of the slope to form a brown-earth (Fig. 4.5) which is less acid than a podzol. On the slope itself the soil tends to be deeper on the gentler gradients than where the slope is steeper. On steep slopes rainwash tends to erode the surface soil and carry it downslope. Also, other types of mass movement (Fig. 1.10) can remove soil from steeper slopes. In extreme cases, a lithosol (azonal soil) can occur in the form of a scree (Fig. 1.13). Screes tend to occur where the slope is broken by free faces (Fig. 1.12). The soil's moisture content tends generally to increase from the summit to the foot of the slope. On the level ground at the foot of the slope water may accumulate in the soil and give rise to gleying, thus producing an intrazonal gley soil (page 161).

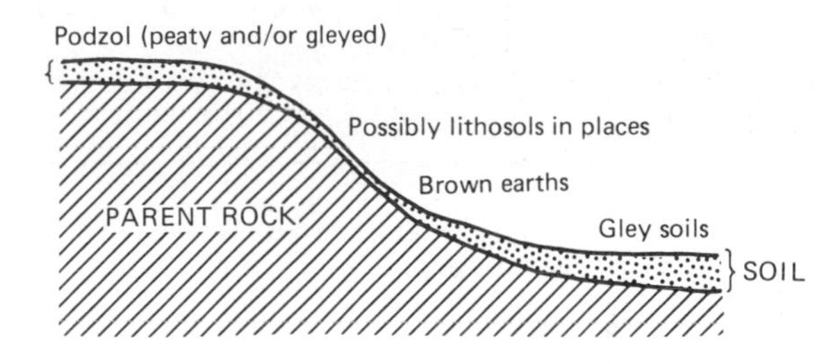

Fig. 4.5 Diagrammatic representation of a soil catena

THE SEVENTH APPROXIMATION

The American Department of Agriculture Soil Survey has introduced a system of soil classification that differs from the zonal system. It classifies soils on the basis of the characteristics of their profiles that have either influenced their development or result from their development. Unlike the zonal system (above), it does not primarily emphasize the influence of climate. Ten major soil groups are identified and then these are divided into a large number of subgroups. These major groups (orders) are listed in Table III.

Table III Seventh Approximation—Major soil orders

Order	Description
1 Entisols	Recent soils. Includes regosols, lithosols and alluvial soils.
2 Vertisols	Mixed or inverted soils.
3 Inceptisols	Young soils with weakly developed profiles.
4 Aridisols	Soils of arid areas including solonchak.
5 Mollisols	Soils rich in humus and bases, including chernozem and rendzina.
6 Spodosols	Leached acid soils, including podzols.
7 Alfisols	Pedalfers with a higher base content, including brown podsolic soils.
8 Ultisols	Deeply weathered ferruginous soils, including some lateritic soils.
9 Oxisols	Very deeply weathered, highly leached soils, including laterites.
10 Histosols	Organic soils including peat and bog soils.

Exercises

1. Explain
   (*a*) the ways in which soil differs from regolith;
   (*b*) why soils differ in colour and texture.
2. (*a*) What factors contribute to the development of a soil profile?
   (*b*) Describe and explain the differences between the profiles of any two types of soil.
3. (*a*) Explain what is meant by the terms 'zonal', 'intrazonal' and 'azonal' when these are applied to types of soil.
   (*b*) Discuss the advantages and the weaknesses of this threefold classification of soils.
4. Describe how and explain why (*a*) the colour, (*b*) the humus content, and (*c*) the direction of movement of soil water, vary so greatly between different types of soil.
5. Why do some soil profiles exhibit clearly marked layers of differing colour, texture, and composition whereas others have more gradual variations.
6. (*a*) Explain what is meant by leaching.
   (*b*) Why does the severity of leaching differ between different types of soil?
7. Describe and explain the characteristics of the soils that you would expect to occur in Britain
   (*a*) in a river flood plain and nearby river terraces;
   (*b*) on a chalk upland;
   (*c*) on hillside slopes;
   (*d*) in a glaciated lowland with sheets of boulder clay and drumlins, kames and eskers.

8. Refer to Figures 4.6 and 4.7.
   (*a*) Name the types of zonal soil that are characteristic of the areas A, B, C, D and E in Figure 4.6.
   (*b*) Explain the relationships between these types of zonal soil and the environmental conditions in which they occur.

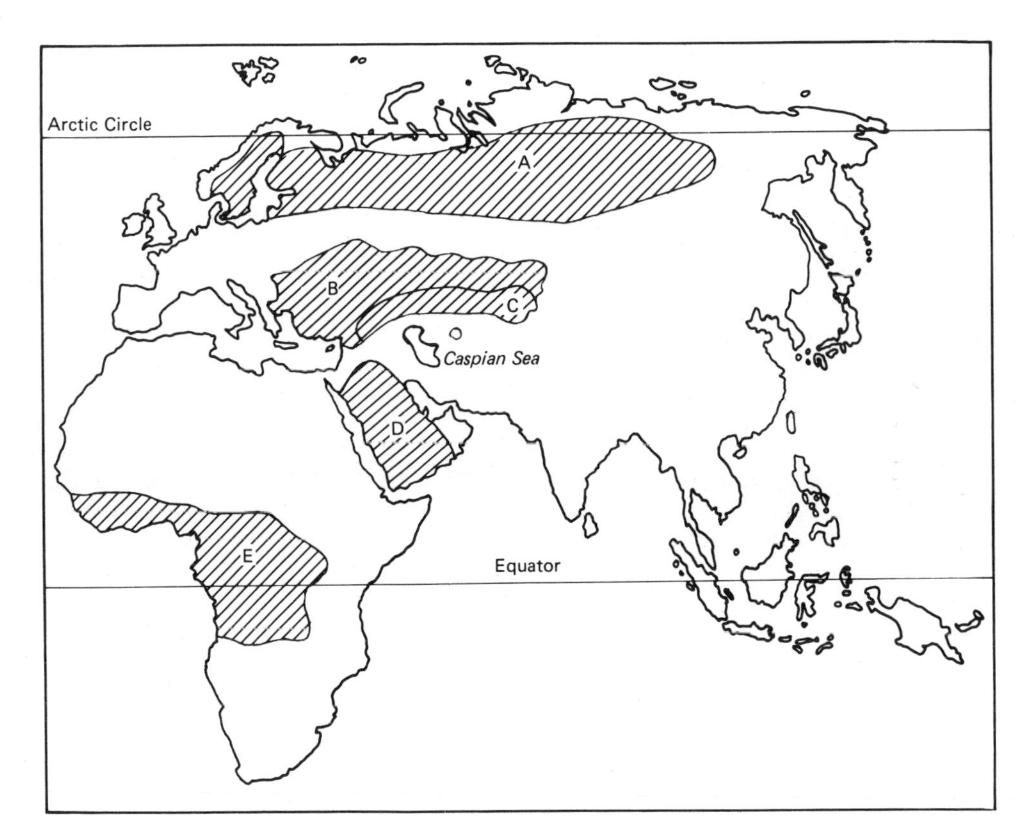

Fig. 4.6 Types of zonal soil in Eurasia and Africa

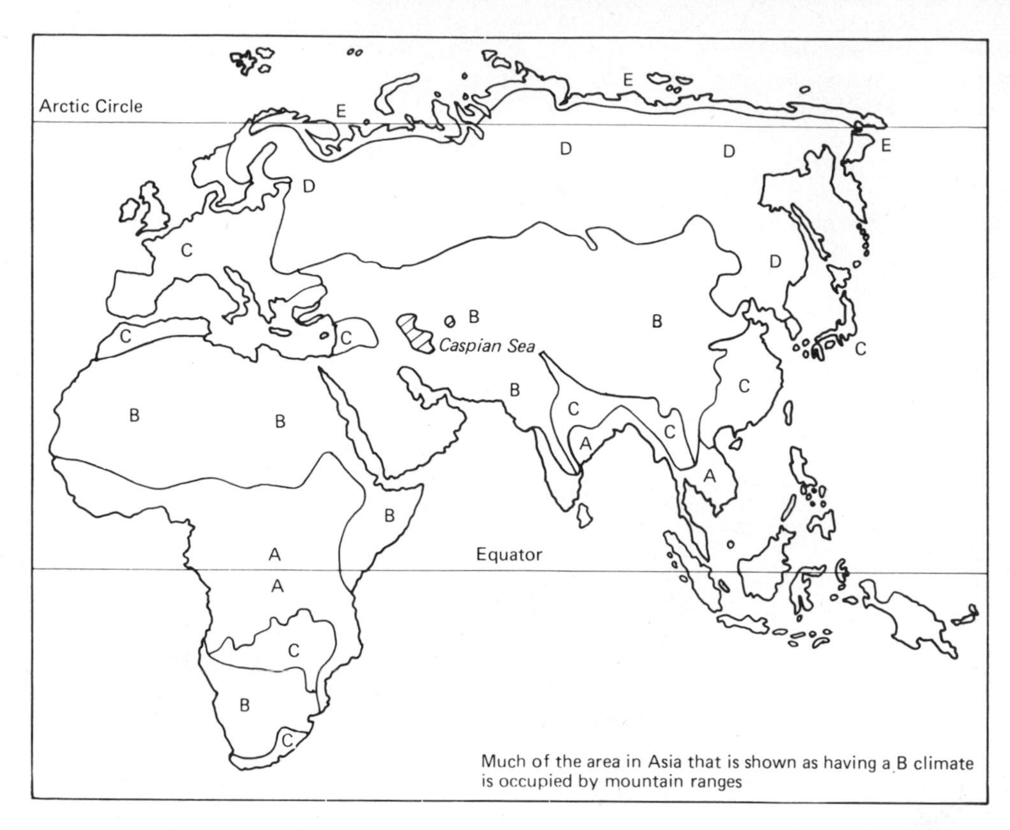

Fig. 4.7 Major climatic regions of Eurasia and Africa (Köppen)

9. The diagrams in Figure 4.8 describe the climate at five different locations.
   (*a*) From the following list identify the type of zonal soil that is associated with each of the climatic types represented by diagrams A to E in Figure 4.8: brown forest soil, chernozem, chestnut-brown soil, podzol, tundra soil.
   (*b*) Choose any three of the soil types listed in (*a*) and explain how their characteristics reflect the influence of climate and vegetation.

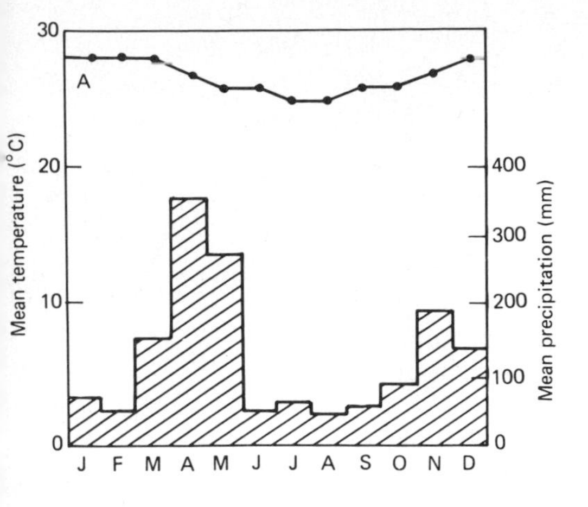

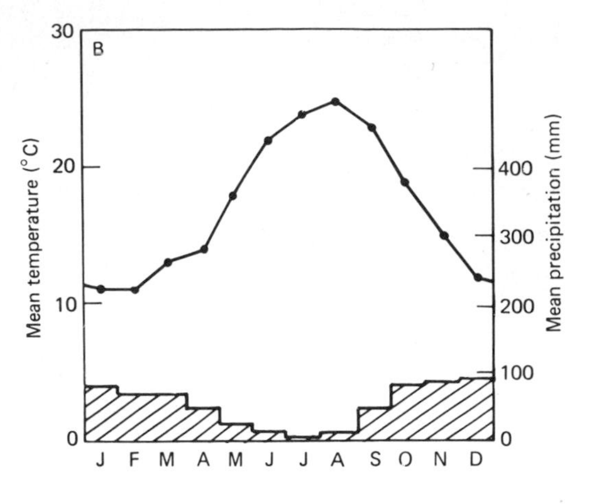

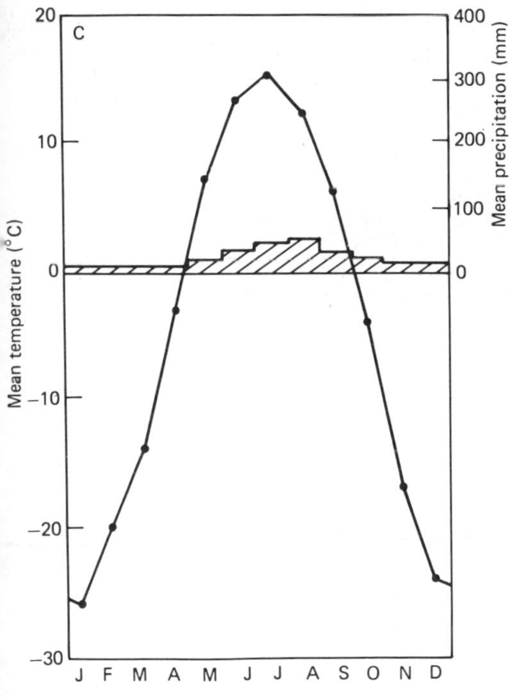

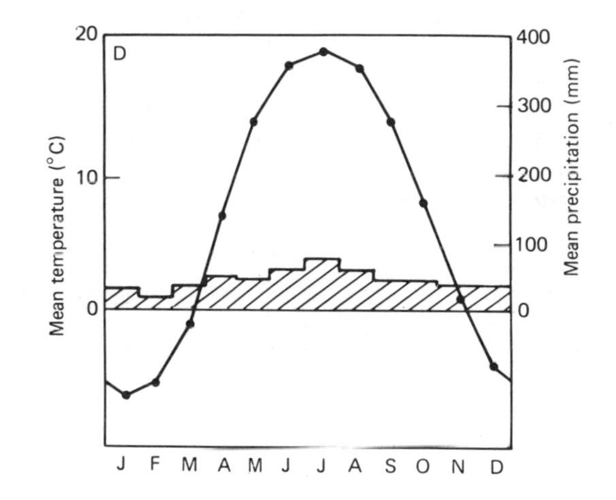

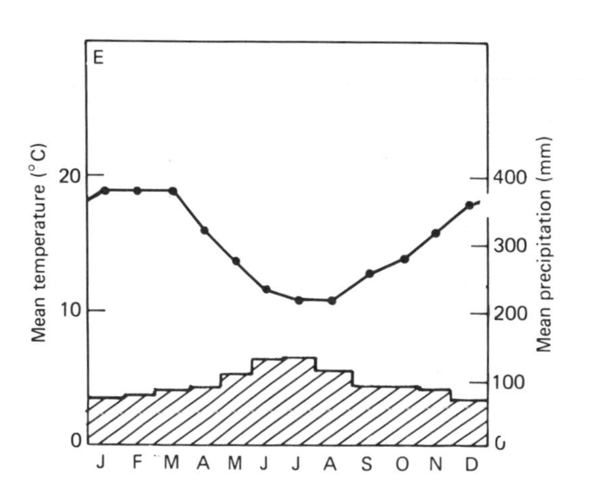

Fig. 4.8 Graphs showing the climate at 5 different locations

Index